FINITE MATHEMATICS
Concepts & Applications

A HORSE   NOT A HORSE?

$$\text{Pr(Arizona} \mid \text{Sweet)} = \frac{\text{Pr(Arizona \& Sweet)}}{\text{Pr(Sweet)}}$$

$$= \frac{20\% \times 95\%}{80\% \times 90\% + 20\% \times 95\%} = \frac{19}{91} \approx 21\%$$

FINITE MATHEMATICS
Concepts & Applications

Tityik Wong
College of Southern Nevada

Prologue

To myself, or whomever it may concern:

What do we want students in our math classes to learn? A high percentage of specific knowledge and skills learned in a mathematics class are never directly used by most students in real life. For example, most students learned how to solve a two variable, two linear equation system at some point during their educational career, but how many of those students actually had to solve such a system in the workplace? I have no solid data to show here, but I would boldly estimate the percentage to be in the neighborhood of 1%%. If you are a mathematics major, then you will use that skill here and there in some subsequent classes you take. If you find a "math-related" job after graduation—some engineering job for example, then you will probably use that skill... never again because if you have to solve a system, it most likely will be a large system and you will use some software to do the job for you. Most people who actually use the skill after graduation are people who make a living teaching that same skill to new students—the math teachers!

How much mathematics should be taught to our students is an ongoing debate, and the issue is too complex to be discussed here.

But while we are still teaching and students are still learning, what do we do?

One of the more prevalent ideas is to teach students "how to learn" or "how to think" (cliché, cliché, but true still). But learning "how to learn" or "how to think" in the absence of content is impossible. Therefore, the selection of topics is of ultimate importance. A topic is chosen to be included in this book, more or less within the traditional finite mathematics framework, when it meets at least one of the following criteria:

1. Immediate applications in our students' lives. This sounds noble but is almost impossible given the wide range of lives our students will lead. So this is reduced to "useful to as many students as possible, I think". For example, almost everyone would benefit from understanding the basic workings of compound interest, I think. Hence the topic "The Mathematics of Finance".

2. Intellectually stimulating and inspiring, and technically manageable for students. For

example, the Gauss-Jordan elimination method for solving a linear system is unlikely to be used by anyone in a real workplace, because large, real life systems are solved by computers. Yet the method itself is so elegant in its design and structure that students will benefit tremendously from solving a few systems by hand. In doing so, the students will get to appreciate how a seemingly formidable task—solving a 500 by 500 linear system for example, can be accomplished by an ingenious method that involves only a few rules. The fact that this can be done will inspire the students to devise simple yet efficient methods when they face difficult situations in real life. This is very important because a great deal of the world's grief has been caused by powerful people trying to solve complex problems with brute force.

3. Interesting and challenging. Most people like to solve challenging problems as long as the problems are not impossible to solve—that's just the way we humans are. Do we have to run marathons in order to survive? No! But people keep doing it just because it is challenging. However, if you all of a sudden require all marathon runners to finish the course in under two hours, then not too many people would run anymore because no one has ever run a marathon in under two hours (there has been no record of such a feat as of the year 2019).

4. Having our "common sense" proven wrong is also a good way to teach us not to jump to conclusions without thorough investigation. The theory of probability and statistics provides ample opportunities for us to do just that. These types of topics will develop the students' critical thinking habits and broaden their intelligence. Probability and statistics are also useful in terms of helping us understand and interpret the vast amount of information we face every day.

We will try our best to use different "real life" stories to reinforce one concept, given that more and more people nowadays have really short attention spans as a result of the nonstop bombardment of information from the digital realm. A teacher feels almost criminal in daring to bring up the concept of "drilling". But drilling is an essential component on the path to the pinnacle of any art, or enlightenment.

We emphasize that mathematics should be placed first, and applications second. If we don't place mathematics first, then we can't do mathematics well. If we can't do mathematics well, then we can't do mathematics applications well. If you play chess well, you will be able to apply some principles in chess to some real-life situations. It is not a good idea though to try to modify the game of chess to make it more closely resemble real life situations because real life situations are too numerous. The best games are the ones that catch some common principles behind a lot of different real life situations.

To the student:

The text itself is very important and should be read with great attention. It is strongly suggested that you have pencils, paper and a calculator available when you read the book so you can immediately verify the computations and other claims made by the author. Mathematics is a brainy art, but it cannot be done without performing a lot of manual work, either. Reading and writing are uniquely human skills. Practise them with great pride!

Exercise problems are an integral part of a textbook. Although efforts have been made to make exercise problems as interesting and challenging as possible, do realize that in order to learn and master any subject, a certain amount of mechanical, repetitive work is absolutely necessary. There is also a gap between "understanding" and "being able to do", and the gap can only be closed through sufficient practice.

Sometimes one technique or method can be used to solve many seemingly unrelated problems; sometimes one problem can be viewed from different angles and solved with different methods. You are encouraged to explore. Do not settle for the correct answer alone.

We will also emphasize the ability to manipulate formulas. Solving problems that involve a lot of letters will develop your abstract thinking ability, which will be beneficial beyond mathematics.

Asking yourself some questions after you have correctly solved a problem is an excellent way to deepen your understanding. Ask yourself how you would solve the problem if certain numbers or parameters were changed.

Each section in this book contains a number of exercise problems that will reinforce the concepts covered in that section.

Some complex problems will be analyzed. These problems are beyond the scope of a traditional finite mathematics textbook, but you will have enough knowledge to follow the solution processes of these problems after finishing the book.

And did I mention you need to read a lot?

Contents

Chapter 1

The Mathematics of Finance

1.1 THE ARITHMETIC SERIES AND SIMPLE INTEREST

Given a number a_1, and a constant d. Add d to a_1 to form the second number a_2, then add d to a_2 to form the third number a_3; continue in this fashion to generate a sequence of numbers. We call such a sequence an *arithmetic sequence*, and the number d the common difference of the sequence.

Example 1.1.1. *Given the first term 3 and the common difference 2, we can generate the arithmetic sequence: 3, 5, 7, 9, 11, ... Notice that from the second term on, every term is the preceding term plus 2:*

$$5 = 3 + 2$$
$$7 = 5 + 2$$
$$9 = 7 + 2$$
$$\vdots$$

Or we can look at it this way:

$$5 = 3 + 2$$
$$7 = 3 + 2 + 2$$
$$9 = 3 + 2 + 2 + 2$$
$$\vdots$$

1

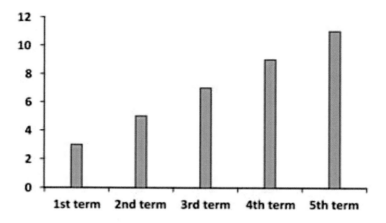

The second term is 3 plus one 2, the third term is 3 plus two 2's, the fourth term is 3 plus three 2's, and so on. We can now see how to find the nth term, it is 3 plus $(n-1)$ 2's, or $3 + (n-1)2$ ◄

In general, given that an arithmetic sequence has first term a_1 and common difference d, the nth term of the sequence is given by the formula:

> The nth term of an arithmetic sequence
>
> $$a_n = a_1 + (n-1)d \qquad \text{or} \qquad a_n = a_i + (n-i)d \qquad\qquad (1.1.1)$$

The sum of some consecutive terms of a sequence is called a *series*. For example, given the arithmetic sequence: 3, 5, 7, 9, 11, ..., The sum of the first five terms $3 + 5 + 7 + 9 + 11$ is a series. The sum of some consecutive terms of an arithmetic sequence is called an *arithmetic series*.

The summation notation, or sigma notation, can be used for a series. For example, the sum of the first 10 terms of the series 3, 5, 7, 9, 11, ...

$$3 + 5 + 7 + 9 + 11 + 13 + 15 + 17 + 19 + 21$$

can be written as

$$\sum_{n=1}^{10} \left[3 + 2(n-1) \right]$$

If you have never seen the summation notation before, here is how it works:

1. The Greek letter sigma, Σ, tells us we are taking a sum.
2. The numbers we are adding together follow the formula $3 + 2(n-1)$, where n is the index of a term. If $n = 1$ then it is the first term, if $n = 2$ then it is the second term, etc. You get the value of the term by substituting the index number for n. The first term is $3 + 2(1-1) = 3$, the second term is $3 + 2(2-1) = 5$, the third term is $3 + 2(3-1) = 7$, and so on.

3. The number following "$n =$" under the letter Σ indicates the starting term. In our example, $n = 1$ tells us the starting term is the first term. The number over the letter Σ indicates the end term in the series. In our example, the number 10 over Σ tells us the end term in the series is the 10th term.

For a general sequence $a_1, a_2, a_3, ...$, the sum of the ith term through the jth term $(i \le j)$ is

$$\sum_{n=i}^{j} a_n = a_i + a_{i+1} + \cdots + a_j$$

For example,

$$\sum_{n=5}^{85} a_n = a_5 + a_6 + \cdots + a_{85}$$

Example 1.1.2. *Find the value of the arithmetic series*

$$3 + 5 + 7 + 9 + 11 + 13 + 15 + 17 + 19 + 21$$

Solution. *We can simply add the numbers together, but that would be very inefficient if there are too many terms in a series. In this situation a formula is more desirable because a formula reduces the amount of work and as a consequence also reduces the chances of making mistakes during computations. Let's label the series S, and write S in two different ways, first with the numbers in ascending order and second with the numbers in descending order:*

$$
\begin{array}{ccccccccccccccccccc}
S & = & 3 & + & 5 & + & 7 & + & 9 & + & 11 & + & 13 & + & 15 & + & 17 & + & 19 & + & 21 \\
S & = & 21 & + & 19 & + & 17 & + & 15 & + & 13 & + & 11 & + & 9 & + & 7 & + & 5 & + & 3
\end{array}
$$

then we add the two equations together, first term to first term, second term to second term, and so on:

$$2S = 24 + 24 + 24 + 24 + 24 + 24 + 24 + 24 + 24 + 24$$

—that's the sum of ten 24's!

$$2S = 10 \times 24$$

which leads to

$$S = \frac{10 \times 24}{2}$$

Notice here that the number 10 in the result indicates that there are 10 terms being added together; the number 24 is the sum of the first term 3 and the last term 21 ◄

Given an arithmetic series $a_1 + a_2 + a_3 + \cdots$, the sum of the tth term through the jth term $(i \le j)$ is given by the formula:

> **The sum of an arithmetic series**
>
> $$\sum_{n=i}^{j} a_n = \frac{(j-i+1)(a_i + a_j)}{2} \qquad \text{in particular} \qquad \sum_{n=1}^{j} a_n = \frac{j(a_1 + a_j)}{2} \qquad (1.1.2)$$

Example 1.1.3. *Given an arithmetic sequence* $100, 102, 104, 106, 108, \ldots$ *Find the sum of the* 3rd *term through the* 57th *term.*

Solution. *The common difference of the sequence is obviously* 2. *The* 3rd *term is* 104, *and the* 57th *term is* $a_{57} = 100 + (57-1)2 = 212$. *The sum of the* 3rd *term through the* 57th *term is*

$$\sum_{n=3}^{57} a_n = \frac{(57-3+1)(104+212)}{2} = 8690 \blacktriangleleft$$

The subject commonly called *Simple Interest* is a direct application of the arithmetic series. Suppose $100 is deposited in an account that carries an interest rate of 6% per year. After one year the principal generates $100 × 6% = $6 of interest, making $106 the balance in the account after one year. Under the simple interest setup, the interest generated will not become part of the principal. If the principal and interest are left in the account and all conditions remain the same, then after another year another $6 of interest will be generated and the balance in the account becomes $112. As we can see the balances in the account after $0, 1, 2, 3, 4, \ldots$ years are $100, $106, $112, $118, $124...—an arithmetic sequence with 100 as the first term and 6 as the common difference.

In general, let P be your principal or initial deposit, i the interest rate per period (typical periods are: a year, a quarter, a month, a week, a day), n the number of periods your money stays in the account, and B the balance in your account after n periods. Under the simple interest setup, the balance is given by the following formula:

> **Balance with simple interest**
>
> $$B = P + Pni \qquad \text{or} \qquad B = P(1 + ni) \qquad (1.1.3)$$

Example 1.1.4. *Suppose an account pays* 18% *simple annual interest, and* $500 *is deposited into the account. If the interest is paid monthly and no money is withdrawn from the account since the initial deposit, find the balance in the account after 30 months.*

Incorrect Solution. $P = 500$, $n = 30$, *and* $i = 18\% = 0.18$. *The balance is therefore*

$$B = 500(1 + 30 \times 0.18) = 3200$$

Correct Solution. *The mistake made above was forgetting to match the units between i and n. If you count the number of periods in months, then you should use the monthly rate; if you count the number of periods in years, then you should use the annual rate. Since the annual rate is 18%, we can convert 30 months to 30/12 = 2.5 years, then use the formula*

$$B = P(1 + ni) = 500(1 + 2.5 \times 0.18) = 725$$

Alternatively, we can use n = 30 months, but then we have to change the annual rate to a monthly rate by dividing the annual rate 18% by 12: i = 18%/12 = 1.5% = 0.015, and

$$B = P(1 + ni) = 500(1 + 30 \times 0.015) = 725 \blacktriangleleft$$

Remark. *The word "month", when used in some mathematical formulas, stands for one-twelfth of one year, which is not exactly the same as a particular month in a year. Suppose there are 365 days in a particular year, then January has 31 days, February has 28 days, and so on. These are real calendar months. But the mathematical month is 365/12 days, or roughly 30.42 days—just when you think math people use precise terms◄*

Example 1.1.5. *I can't resist the new, cool, big, curved screen TV set but it sells for a price of $3000, which I can't afford. The store allows me to have the TV but wants to charge me 24% simple annual interest and I have to pay the principal plus interest off in fifteen equal monthly payments. How much is my monthly payment?*

Solution. *We need to find out how much is the principal plus interest, then divide that amount by 15 to get the monthly payment amount. Since it is done under the simple interest assumption, the principal plus interest is*

$$B = P(1 + ni) = 3000\left(1 + \frac{15}{12} \times 0.24\right) = 3900$$

and the monthly payment amount is
$$\frac{3900}{15} = 260 \blacktriangleleft$$

– EXERCISES 1.1 –

1. The first three numbers in an arithmetic sequence are 2.1, 1.8, and 1.5. Find the fourth number.

2. The first term of an arithmetic sequence is 5, the common difference is 0.8. Find the 150th term in the sequence.

3. The 5th term of an arithmetic sequence is 4, the common difference is 2.5. Find the 13th term.

4. The first term and the sixth term of an arithmetic sequence are 4 and 6, respectively. Find the common difference. $\frac{6-4}{6-1} = 2.5 = \boxed{.4}$

5. The 32nd term of an arithmetic sequence is 18.2, the common difference is −1.2. Find the 13th term. $32 - 13 = 19$ $18.2 + (1.2)(19_{\overline{18}}1)$ $18.2 + 21.6$ $a_n = a_1 + (n-1)d$

6. The first term of an arithmetic sequence is 5, the common difference is 0.8. Find the sum of the first 200 terms.

7. The sum of the first 30 terms of an arithmetic sequence is 1350. The first term is 1.5. Find the common difference.

8. Divide one year into 12 equal periods and call each period a "month". Convert the following time in months to time in years.

    (a) 12 months.   (c) 18 months.   (e) 27 months.   (g) 10 months.
    (b) 24 months.   (d) 15 months.   (f) 30 months.

9. Suppose an account pays 14.4% simple annual interest, and $6000 is deposited into the account. If the interest is paid monthly and no money is withdrawn from the account since the initial deposit, find the balance in the account after (assume there are 365 days in a year)

    (a) Two years.   (b) 36 months.   (c) 42 months.   (d) 1533 days.

10. Suppose I need to borrow $2000 from my neighbor The Saver. The Saver charges 36% simple annual interest rate and I have to pay the principal plus interest off in eighteen equal monthly payments.

    (a) How much will be the interest charge?

    (b) How much will be my monthly payment?

11. The first two terms of an arithmetic sequence are 2.5 and 3.2. Find the sum of the first 700 terms.

12. Afondi borrowed 100 olives from Dogglass, and needs to pay 105 olives to Dogglass after one month. What is the annual interest rate Dogglass is charging Afondi?

1.2 THE GEOMETRIC SERIES AND COMPOUND INTEREST

Given a number a_1, and a constant r. Multiply a_1 with r to form the second number a_2, then multiply a_2 with r to form the third number a_3; continue in this fashion to generate a sequence of numbers. We call such a sequence a *geometric sequence*, and the number r the common ratio of the sequence.

Example 1.2.1. *Given the first term 3 and the common ratio 2, we can generate the arithmetic sequence: 3, 6, 12, 24, 48, ... Notice that from the second term on, every term is the preceding term times 2:*

$$6 = 3 \cdot 2$$
$$12 = 6 \cdot 2$$
$$24 = 12 \cdot 2$$
$$\vdots$$

Or we can look at it this way:

$$6 = 3 \cdot 2^1$$
$$12 = 3 \cdot 2^2$$
$$24 = 3 \cdot 2^3$$
$$\vdots$$

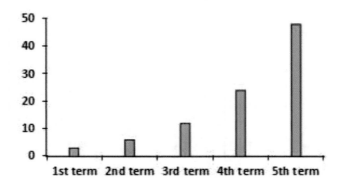

The second term is 3 times 2, the third term is 3 times 2 squared, the fourth term is 3 times 2 cubed, and so on. We can now see how to find the nth term, it is 3 times 2 to the $(n-1)^{th}$ power, or $3 \cdot 2^{n-1}$ ◄

In general, given that a geometric sequence has first term a_1 and common ratio r, the nth term of the sequence is given by the formula:

The nth term of a geometric sequence

$$a_n = a_1 r^{n-1} \quad \text{or} \quad a_n = a_i r^{n-i} \tag{1.2.1}$$

Example 1.2.2. *Suppose the first term of a geometric sequence is* 500 *and the common ratio is* 0.8. *Find the* 7th *term of the sequence.*

Solution.

$$a_7 = a_1 r^{7-1} = 500 \cdot 0.8^6 = 131.072 \blacktriangleleft$$

What if we want to add some consecutive terms in a geometric sequence together (such a sum is, as you might have guessed, called a *geometric series*)? There is of course a formula we can use, but the derivation of the formula is so awesome we can't resist going through it with an example.

Example 1.2.3. *Find the value of the geometric series*

$$5 + 15 + 45 + 135 + 405$$

Solution. *Let the sum be S:*

$$S = 5 + 15 + 45 + 135 + 405$$

It is easy to see that the common ratio is 3 *(we divide the second term* 15 *by the first term* 5 *to find the ratio). The term after* 405 *is then* $405 \times 3 = 1215$. *Let's add* 1215 *to both sides of the last equation:*

$$S + 1215 = 5 + 15 + 45 + 135 + 405 + 1215$$

On the right-hand side of the equation we factor out a common factor 3 *among the last 5 terms:*

$$S + 1215 = 5 + 3(5 + 15 + 45 + 135 + 405)$$

The part on the right-hand side inside the parentheses is S itself. Now we have

$$S + 1215 = 5 + 3S$$

Solving for S we get

$$2S = 1215 - 5$$

or

$$S = \frac{1215 - 5}{2}$$

Let's look at the numbers: 1215 *is the* 6th *terms in the series, according to formula* (1.2.1), $1215 = 5 \cdot 3^{6-1}$. *The* 5 *in the final answer is the* 1st *term. The* 2 *in the final answer comes from* $3S - S = 2S$, *where* 3 *is the common ratio. So* 2 *actually comes from* (*common ratio* − 1). *Now we can see that*

$$S = \frac{5 \cdot 3^5 - 5}{3 - 1} = \frac{5(3^5 - 1)}{3 - 1} \blacktriangleleft$$

Below is the general formula:

> **The sum of a geometric series**
>
> $$\sum_{n=i}^{j} a_n = \frac{a_i(r^{j-i+1} - 1)}{r - 1} \quad \text{in particular} \quad \sum_{n=1}^{j} a_n = \frac{a_1(r^j - 1)}{r - 1} \qquad (1.2.2)$$

Example 1.2.4. *Given a geometric sequence* 100, 101, 102.01, 103.0301, ..., *find the sum of the* 1st *term through the* 15th *term.*

Solution. *The common ratio* r *can be found by dividing the second term by the first term (or the third term by the second term):* $r = 101/100 = 1.01$. *Using formula* (1.2.2)

$$\sum_{n=1}^{15} a_n = \frac{a_1(r^{15} - 1)}{r - 1} = \frac{100(1.01^{15} - 1)}{1.01 - 1} \approx 1609.69 \blacktriangleleft$$

In the last section we see that under the simple interest setup, the interest earned is not considered part of the principal. If this rule is changed and every time interest is calculated it is immediately added to and becomes part of the principal for the next period, we have what is called the *compound interest* setup.

Suppose $100 is deposited in an account that carries an interest rate of 6% per year. After one year the principal generates $100 × 6% = $6 of interest. Under the compound interest setup, the interest generated becomes part of the principal, making $106 the principal for the second year. If the principal is left in the account and all conditions remain the same, then after another year $106 × 6% = $6.36 of interest will be generated and the balance in the account becomes $112.36. We can verify that the balances in the account after 0, 1, 2, 3, 4... years are $100, $106, $112.36, $119.1016, $126.247696, ...—a geometric sequence with 100 as the first term and 1.06 as the common ratio.

In general, let P be the principal or initial deposit, i the interest rate per period, n the number of periods your money stays in the account, and B the balance in your account after n periods. Under the compound interest setup, the balance is given by the following formula:

> **Balance with compound interest**
>
> $$B = P(1 + i)^n \qquad\qquad (1.2.3)$$

Example 1.2.5. *$3000 is deposited into an account that pays 6% annual interest with interest compounded monthly. What is the balance of the account at the end of the 5th year?*

Solution. *When using formula (1.2.3), the interest rate and the number of compounding periods must have matching units. Since interest is compounded monthly, the interest rate must be a monthly rate. The annual rate is 6%, so the monthly rate is $6\%/12 = 0.5\% = 0.005$. Also notice that there are $5 \times 12 = 60$ periods because there are 12 months in a year and the money is in the account for five years. According to (1.2.3), the balance after five years is*

$$B = 3000(1 + 0.005)^{60} \approx 4046.55 \blacktriangleleft$$

Example 1.2.6. *An investment will pay an annual interest of 5% with interest compounded quarterly. How much should be invested now in order to have a balance of $5000 in the account after three years?*

Solution. *The compounding period is a quarter, or three months. There are four quarters in a year, so the quarterly rate is $5\%/4 = 1.25\% = 0.0125$. We want the balance to be 5000 after 12 periods (3 years times 4 periods per year), and we want to know the principal. (1.2.3) gives us*

$$5000 = P(1 + 0.0125)^{12}$$

or

$$5000 = P(1.0125)^{12}$$

So

$$P = 5000(1.0125)^{-12} \approx 4307.54$$

Answer: $4307.54 should be invested now in order to have $5000 in three years ◀

– EXERCISES 1.2 –

1. The first two terms of a geometric sequence are 2 and 3. Find the third term and the fourth term.

2. The first term of a geometric sequence is 100, the common ratio is 1.02.

  (a) Find the 20th term.

  (b) Find the sum of the first 20 terms.

3. The fifth term of a geometric sequence is 8, the common ratio is 0.25.

  (a) Find the third term.

  (b) Find the sum of the third term through fifteenth term.

4. Suppose a savings account pays 5% annual interest. $500 is deposited into the account. Find the balance in the account after two years if

  (a) The interest is compounded annually.

  (b) The interest is compounded monthly.

  (c) Observe that the balance is slightly higher when interest is compounded monthly compared to annually. What is the reason?

5. J.J. needs $500 in fifteen days to pay off his debt. His brother has an investment opportunity that pays a whopping 990% annual rate of return, with interest compounded daily. How much should J.J. give his brother today in order to have $500 in fifteen days?

6. Ami bought a house five years ago for $200,000. Today the house is worth $300,000. Estimate the annual rate of increase of the value of the house over the past five years.

7. APY is short for Annual Percentage Yield. APY is the amount of interest one earns if one invests $1 in an account that pays a certain APR (Annual Percentage Rate) over one year. So how is APY different from APR? $APY = APR$ if the interest is compounded annually. If the interest is compounded more than once per year, then APY will be slightly higher than APR. The compound interest formula $B = P(1 + i)^n$ is all we need to figure out APY. For example, if $APR = 6.8\%$ with interest compounded daily, then $i = 0.068/365 = 0.000186301$, $n = 365$, and remember $p = 1$. So the balance is $B = 1(1 + 0.000186301)^{365} \approx 1.07$. The investment is $1 and the interest amount is $0.07. The APY is therefore 0.07, or 7%.

8. Suppose an account pays a 9% annual interest with interest compounded monthly. Find its APY.

9. Suppose a bank offers a 12-month CD with $APY = 2\%$ and $APR = 1.98\%$, then the interest must be compounded

  (A) Monthly   (B) Quarterly   (C) Weekly   (D) Daily

1.3   SIMPLE INTEREST VS. COMPOUND INTEREST

You should now write down formulas (1.1.3) and (1.2.3) side by side, and stare at them intensely for a while. They look equally easy, slightly different. But how are they different?

In mathematical terms, we call (1.1.3) a linear growth model, and (1.2.3) an exponential growth model. In (1.1.3), the values of B at $n = 0, 1, 2, 3, \ldots$ fall on a straight line; while in (1.2.3) the values of B at $n = 0, 1, 2, 3, \ldots$ fall on an exponential curve. Let's look at an example.

Example 1.3.1. *Suppose $P = 100$, $i = 20\%$.*

(a) *Use formula (1.1.3) to find the values of B at $n = 0, 1, 2, 3, 4, 5, 6, 7, 8, 9$.*
(b) *Use formula (1.2.3) to find the values of B at $n = 0, 1, 2, 3, 4, 5, 6, 7, 8, 9$.*
(c) *Plot the B values above in the same coordinate system to show how the balances grow differently under the simple interest and compound interest setups.*

Solution.

(a) *Under the simple interest setup,*

$$B = P(1 + ni) = 100(1 + 0.20n) = 100(1.20n)$$

Plugging in $0, 1, 2, \ldots, 9$ for n we find that the B values (rounded to the hundredth place) are:
100, 120, 140, 160, 180, 200, 220, 240, 260, 280.

(b) *Under the compound interest setup,*

$$B = P(1 + i)^n = 100(1 + 0.20)^n = 100(1.20)^n$$

Plugging in $0, 1, 2, \ldots, 9$ for n we find that the B values (rounded to the hundredth place) are:
100, 120, 144, 173, 207, 249, 299, 358, 430, 516.

(c) *For ease of reading, let's put the numbers above in a table first then graph them.*

n	0	1	2	3	4	5	6	7	8	9
B (Simple)	100	120	140	160	180	200	220	240	260	280
B (Compound)	100	120	144	173	207	249	299	358	430	516

The graph below shows that the balances under the simple interest setup fall on a straight line, while the balances under the compound interest setup fall on an exponential curve whose growth accelerates as the number of periods increases.

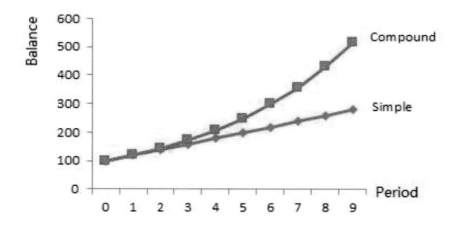

The two curves above are relatively close when n is small, but they start to split as n grows large. The balance in the compound interest account grows faster and faster over time. The following example further demonstrates this point.

Example 1.3.2. *Suppose $P = 100$, $i = 20\%$.*

   (a) *Use formula (1.1.3) to find the value of B when $n = 30$.*
   (b) *Use formula (1.2.3) to find the value of B when $n = 30$.*

Solution.

   (a) $B = P(1 + ni) = 100(1 + 30 \cdot 0.20) = 700$
   (b) $B = P(1 + i)^n = 100(1 + 0.20)^{30} = 23737.63$—*almost 3300% more than the $700 balance based on simple interest*◄

– EXERCISES 1.3 –

Mathematics equations don't always have nice, neat solutions as most textbook examples show. Some problems are so hard they cannot be solved—those of course won't be discussed here. Some problems can be solved by brute force, or some creativity, or both. We like problems of this nature because they force us to really understand everything about the problem we are facing, rather than grab a formula, make some substitutions and computations, get an answer, and call it a day.

   1. Professor Gulliver has retired from CSN after 50 years of excellent service. He still teaches one class per semester just because. As a reward for his contribution to the college, he

15 years

Trial + error using diff years in place of n in compound formula

is paid $2000 per credit hour per semester he teaches, guaranteed forever. A part time teacher is paid $1000 per credit hour per semester. However, a part time teacher's pay is increased by 5% per year. How many years will it take for a part time teacher's pay to surpass Professor Gulliver's pay?

2. I invest $5000 in some place that pays 5% simple annual interest, and at the same time you invest $4000 in some other place that pays 5% annual interest but with interest compounded annually. Suppose both of us never withdraw money from our respective investments. How many years will it take for your investment to have a higher balance than my investment for the first time?

    (a) Why does the question even make sense if both investments pay 5% and I invest more than you do? How can your balance surpass mine?

    (b) Solve this problem by guessing, scientifically. First use a relatively small number (3 years, for example), find the balance of my investment and the balance of your investment—the goal here is to have my balance higher than your balance. Then use a rather large number (30 years, for example), and find the balance of my investment and the balance of your investment again—the goal here is to have your balance higher than my balance. Now pick a time between 3 years and 30 years and do it again. Meditate on this for a while until you see the way to a solution. It will still take some work, but once you see the path, you will be so thrilled you won't mind the work—well, hopefully you won't.

    (c) Solve the problem by using a graphing calculator or computer software to graph both $y = 5(1 + 0.05x)$ and $y = 4(1 + 0.05)^x$ in the same coordinate system. Find out where the two lines intersect. The x-coordinate of their intersection point gives us the answer.

    (d) How and why does the graphing method described above work? Why is it OK to change 5000 and 4000 to 5 and 4, respectively?

3. Suppose in the first problem above everything else stays the same but Professor Gulliver gets a 2% COLA (cost of living adjustment) per annum. How many years will it take for a part time teacher's pay to surpass Professor Gulliver's pay?

4. The growth formulas become decay formulas if the growth rate is negative. That's the beauty of mathematics. We simply use a negative rate and go about business as usual. Suppose you use your car for business, and need to calculate your car's value for tax filing purposes. The government may use a linear depreciation formula because it is easier for everyone. Assume a new car has a life of 8 years, then after 8 years it is worth $0. If the car's value depreciates the same amount every year, then the depreciation rate is $-100\%/8 = -12.5\%$ of the new car's value. If a new car is worth $30000, how much is it worth after 3 years? Since the depreciation is linear, the simple interest formula applies

here, with $P = 30000$, $n = 3$, and $i = -0.125$. We find the value of the car after 3 years to be $B = P(1 + ni) = 30000(1 + 3 \cdot (-0.125)) = 18750$. Oh, yes, no work required here on your part, except for reading, which is actually some serious work, if you are honest.

5. In the real world, the depreciation of a car's value might follow more closely an exponential curve with a negative rate. Because most of the time a car is never worth $0, even when it reaches retirement age. If a car loses 12.5% of its CURRENT value (note that it is not 12.5% of its initial value) in the following year, how much is it worth after three years if it is worth $30000 now? In this case we use the compound interest formula $B = P(1 + i)^n$ with $P = 30000$, $n = 3$, and $i = -0.125$.

6. Use the same numbers in the previous problem. How much is the car worth after 20 years? 30 years?

1.4 ADDITIONAL EXAMPLES AND FORMULAS

A financial calculator can do all the common financial calculations. But what we are after here is "mathematical senses"—we want to be able to appreciate the why and the how, and in the process develop a deeper understanding of a mathematical formula.

Example 1.4.1. *Mr. and Mrs. Smith decide to open a retirement account on January 1 of the coming year. They will deposit $8000 into the account at the beginning of each year for the next 15 years. Suppose the account pays an annual interest of 7% with interest compounded annually. Find the balance in the account right after they have made the 15th deposit.*

Solution. *One way to find the answer is by brute force. After one year the account will have a balance of $8000(1 + 0.07) = 8560$. Now another $8000 is added to the account so the account starts the second year with a balance of $16560. At the end of the second year the balance becomes $16560(1 + 0.07) = 17719.20$, and another $8000 is added to the account at the beginning of the 3rd year. Continue in this fashion until we reach the 15th deposit.*

The method above is highly undesirable mathematically for a couple of reasons: (a) It is laborious and hence prone to human errors. (b) We have to go through the same process every time a minor change is made to the situation—for example, when the deposit amount is changed, or the interest rate is changed.

We want to find the answer of course, but we are also a lot greedier than that. We want to find a formula that can be used with any deposit amount, any interest rate, and any number of compounding periods.

We have seen how clumsy it would be if we used the brute force method. Here is a different way to

* compound interest is better for paying off loan *

look at the situation: We could treat the account as 15 *accounts combined—every time a deposit is made, we imagine a new account is opened. There are* 14 *years between the first deposit and the fifteenth deposit, so the first deposit has* 14 *years to grow, and the final balance can be computed by formula* (1.2.3). *Similarly, the second deposit has* 13 *years to grow, the third deposit has* 12 *years to grow, and so on. We list below the balances of the* 15 *deposits, the sum of which will be the total balance in the account after the* 15th *deposit is made.*

$$\text{Balance of first deposit after } 14 \text{ years:} \quad 8000(1.07)^{14}$$
$$\text{Balance of second deposit after } 13 \text{ years:} \quad 8000(1.07)^{13}$$
$$\vdots$$
$$\text{Balance of thirteenth deposit after } 2 \text{ years:} \quad 8000(1.07)^{2}$$
$$\text{Balance of fourteenth deposit after } 1 \text{ year:} \quad 8000(1.07)$$
$$\text{Balance of fifteenth deposit after } 0 \text{ years:} \quad 8000$$

The numbers $8000, 8000(1.07), 8000(1.07)^2, ..., 8000(1.07)^{14}$ *form a geometric sequence where the first term is* 8000 *and the common ratio is* 1.07. *The sum of all the terms, according to formula* (1.2.2), *is*

$$8000\frac{1.07^{15} - 1}{1.07 - 1} \approx 201032.18 \blacktriangleleft$$

Example 1.4.2. *Using the last example as a guide, compute the balance in an account after the* 20th *deposit if we deposit* $2000 *at the beginning of every year and the annual interest rate is* 5% *with interest compounded annually.*

Solution. *The balance is*

$$2000\frac{1.05^{20} - 1}{1.05 - 1} \approx 66131.91$$

Note that if the interest rate is zero, the balance is simply $2000 \times 20 = 40000$, *almost* 40% *less than* 66131.91 $\blacktriangleleft$

Example 1.4.3. *Suppose* $30 *is deposited into an account at the beginning of every month and the annual interest rate is* 6% *with interest compounded monthly. Find the balance after the* 60th *deposit.*

Solution. *Since a deposit is made every month and interest is compounded monthly, we first convert the annual rate* 6% *to a monthly rate of* 0.5%. *The total number of deposits is* 60, *the balance is*

$$30\frac{1.005^{60} - 1}{1.005 - 1} \approx 2093.10$$

That, by the way, is how much you can save if you save $1 *a day for* 5 *years* $\blacktriangleleft$

Let's now look at how borrowing and paying a loan work.

Under the simple interest setup, computations are a lot easier and we will demonstrate the point with the following example.

Example 1.4.4. *I borrowed* $2000 *at a simple interest rate of* 18% *and agreed to pay the loan off in* 5 *years with* 60 *equal monthly payments. How much is my monthly payment?*

Solution. Since it is under the simple interest setup, the balance is, according to formula (1.1.3),

$$2000(1 + 0.18 \times 5) = 3800$$

There are 60 *months in* 5 *years so the monthly payment amount is*

$$\frac{3800}{60} \approx 63.33 \blacktriangleleft$$

If we simply move on at this point, we are doing ourselves a disservice. Let's look at the numbers in the last example, and ponder some issues here.

If there is no interest charge, a $2000 loan can be paid off by making 60 equal monthly payments of about $33.33 each. At 18% annual rate, the payment amount becomes $63.33, that's almost double of $33.33. Isn't the interest rate only 18%? How did this happen?

It turns out the only good thing about the simple interest setup is the ease of calculating the payment amount. Other than that it is extremely unfair, or at least misleading. Under the simple interest setup, the borrower is assumed to owe the lender the amount of the loan plus the interest the moment the loan is taken out. The borrower then pays the lender a fixed amount at certain time interval until the sum is paid off.

The fairer method is to acknowledge that every time a payment is made, the balance is brought down immediately, then goes up again according to the agreed upon rate until the next payment is made. We will use the numbers from the last example to demonstrate how this works.

Start with a $2000 loan, with 18% annual interest rate.

The balance after one month is $2000 + 2000 \times \frac{0.18}{12} = 2030$. At this point a payment of $63.33 is made, bringing the balance down to $1966.67. This is also the beginning balance for the second month.

The balance after two months is $1966.67 + 1966.67 \times \frac{0.18}{12} = 1996.17$. After another payment of $63.33 is made, the balance becomes $1932.84.

We can see how this progresses now: Over one month we increase the current balance according to the agreed upon rate, then bring the balance down immediately by the payment amount after we make a payment.

The balance after the third payment is made at the end of the third month is $1898.50.

The balance after the fourth payment is made is $1863.65.

The balance after the fifth payment is made is $1828.27.

You can verify that the balance after 43rd payment is made is roughly $7.

Actually, I hope you didn't go through all the calculations to verify my last claim. We will find a formula later. For now let's examine the numbers again: the loan is almost paid off after 43 payments of $63.33 each, that's 3 years and 7 months. Remember under the simple interest setup, it takes 5 years to pay off the loan at $63.33 per month. Right here we can see the "unfairness" of the simple interest setup.

The scenario above is similar to when you take out a fixed interest loan to buy a house or a car. For example, if I am buying a car and I borrow $15000 for three years at 9% annual rate, I will have to pay $477 per month for 36 months to pay off the loan. How do we know it is $477 per month?

We want the balance to be zero after 36 payments. Assume the monthly payment amount is M. Since the monthly interest rate is $9\%/12 = 0.75\%$, every month we multiply the balance by 1.0075, then immediately subtract from it the monthly payment amount M.

Time	Balance
Start	15000
After the 1st payment	$15000(1.0075) - M$
After the 2nd payment	$(15000(1.0075) - M)(1.0075) - M$
	$= 15000(1.0075)^2 - 1.0075M - M$
After the 3rd payment	$(15000(1.0075)^2 - 1.0075M - M)(1.0075) - M$
	$= 15000(1.0075)^3 - 1.0075^2 M - 1.0075M - M$
$\vdots$	$\vdots$
After the 36th payment	$15000(1.0075)^{36} - 1.0075^{35}M - 1.0075^{34}M - \cdots - 1.0075M - M$

In the balance after the 36th payment, the sum of the second term through the last term is a geometric series whose sum can be found by formula (1.2.2):

Balance after the 36th payment

$$= 15000(1.0075)^{36} - 1.0075^{35}M - 1.0075^{34}M - \cdots - 1.0075M - M$$

$$= 15000(1.0075)^{36} - \left(\frac{1.0075^{36} - 1}{1.0075 - 1}\right)M$$

$$= 15000(1.0075)^{36} - \left(\frac{1.0075^{36} - 1}{0.0075}\right)M$$

Since we want this balance to be zero, we have

$$15000(1.0075)^{36} - \left(\frac{1.0075^{36} - 1}{0.0075}\right)M = 0$$

and so

$$M = \frac{0.0075}{1 - (1.0075)^{-36}} 15000 \approx 477.00$$

Suppose a loan in the amount of P is lent to the borrower at an interest rate of i per period (here a period is an agreed-upon fixed time interval such as 30 days or 14 days), and interest is compounded at the end of each period. If the loan is to be paid off in n equal payments, one payment per period, then the amount per payment is given by the following formula:

Amount per payment, over n periods, with interest compounded at the end of each period

✳ not simple

$$M = \frac{i}{1 - (1 + i)^{-n}}P \tag{1.4.1}$$

The next example compares the amount per payment of a loan under the simple interest setup, and the amount per payment under the compound interest setup (formula (1.4.1)).

Example 1.4.5. *Suppose a $1000 loan carries a 25% annual interest rate, and is to be paid off in five years with five equal payments.*

(a) *Find the amount per payment under the simple interest setup.* $B = P \cdot (1 + ni)$

(b) *Find the amount per payment under the compound interest setup using formula (1.4.1).*

Solution.

(a) *Principal + Interest* $= 1000 + 1000 \times 0.25 \times 5 = 2250$. *So amount per payment is*

$$\frac{2250}{5} = 450$$

(b) *Use formula (1.4.1) with* $P = 1000$, $i = 0.25$, *and* $n = 5$, *we find the amount per payment to be*

$$M = \frac{0.25}{1 - (1 + 0.25)^{-5}} 1000 \approx 371.85$$

which is about 17% less than the amount under the simple interest setup.

The following graph shows the two balances of the two loans, right after every payment is made, through their respective five year terms. Under the simple interest setup, the beginning balance is Principal + Interest = $2250, the balance is then reduced by $450 every year for five years. The balances follow a straight line from start to finish. Under the compound interest setup, the beginning balance is the principal $1000. The first year balance is computed by increasing the balance by 25% first, then reducing it by the payment amount of $371.85: $1000 × 1.25 − $371.83 = 878.15. We can see that the balances follow a curve that dips more and more steeply as time progresses, meaning that the closer it is towards the end of the term, the more the balance is reduced after each payment.

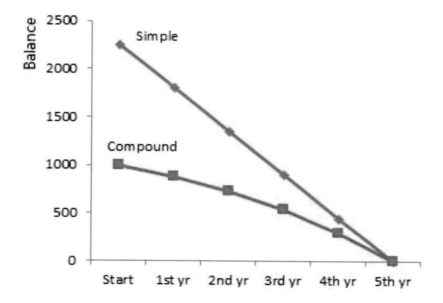

The next graph shows the balance of the loan before and after each payment under the compound interest setup.

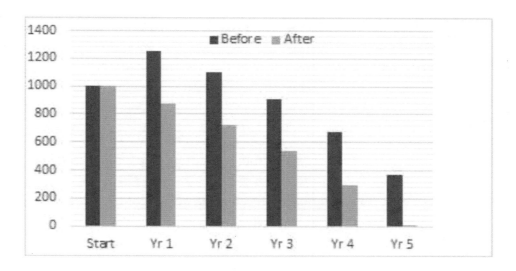

Starting from Year 1, the Before bar is the previous After bar times 1.25, and the After bar is the Before bar minus 371.85. At Year 5, the Before bar is exactly 371.85, so the After bar is zero after subtracting 371.85 from the Before bar. The After bars form the Compound curve in the previous graph◀

– EXERCISES 1.4 –

1. Use the first three examples in the section as a model. Find the balance in an account after the 10th deposit if we deposit \$3000 at the beginning of every year and the annual interest rate is 4% with interest compounded annually.

2. Use the same numbers from the previous problem. Find the balance at the end of the 10th year.

3. Find the balance in an account after the 90th deposit if we deposit \$200 at the beginning of every month and the annual interest rate is 3% with interest compounded monthly.

4. Bart took out a \$500,000 loan to buy a yacht. The loan had a fixed, beautiful annual interest rate of 9%. Suppose Bart wanted to pay off the loan in 10 years with equal monthly payments = Compound

   (a) How much would be the amount per payment?

   (b) How much interest would Bart have paid after the loan was paid off?

5. Use the same story in the previous problem. Instead of a monthly payment schedule, suppose Bart wanted a biweekly payment schedule and he wanted to pay off the loan in 480 weeks (a little over 9 years and 2 months, 240 payments total).

   (a) How much would be the amount per payment?

   (b) How much interest would Bart have paid after the loan was paid off?

6. In this problem we try to instill in you some permanent wisdom (that, by the way, cannot be found in any other finite math textbooks). Suppose you make a $1,000,000 investment. Find the return of your investment after three years in each of the following scenarios. For ease of computation, assume that interest is compounded annually.

   (a) The first-year return rate is 20%, the second 40%, and the third 60%.

   (b) The first-year return rate is 20%, the second 60%, and the third 40%.

   (c) The first-year return rate is 40%, the second 20%, and the third 60%.

   (d) The first-year return rate is 40%, the second 60%, and the third 20%.

   (e) The first-year return rate is 60%, the second 20%, and the third 40%.

   (f) The first-year return rate is 60%, the second 40%, and the third 20%.

   (g) The return rate is 40% in all three years.

   (h) The arithmetic average of 20%, 40% and 60% is 40%. The results above should tell you how consistency beats excitement in long-term investments.

Chapter 2

Sets

2.1 THE SET LANGUAGE

Set theory has broad applications in mathematics, including probability theory and logic, among others.

A *set* is a collection of distinct objects. The objects are called the *elements* or *members* of the set.

The name of a set is normally denoted by an uppercase letter from the beginning of the English alphabet: A, B, C and so on. The elements in a set are enclosed in a pair of cursive brackets if they are not too large a group, or described in plain language. The latter form can also be written in a fancier manner called the set-builder notation. Suppose a set contains the elements 2, 4, 6, 8. If we name this set A, then A can be described in one of the following three ways:

1. Roster notation: $A = \{2, 4, 6, 8\}$
2. Description notation: $A = \{$all even numbers between 1 and 9$\}$
3. Set-builder notation: $A = \{x | x$ is an even number between 1 and 9$\}$

You see the set-builder notation is essentially the same as the description notation, but designed to look a little fancier and a little bit more academic.

The *empty set* is the set that contains nothing, and is denoted by the symbol Ø, or { }.

For a problem under consideration, the *universal set* is the set that contains all elements of interest. For example, if we are talking about days of the week, then the universal set contains seven elements: {Sunday, Monday, Tuesday, Wednesday, Thursday, Friday, Saturday}. The universal set can be denoted by the uppercase letter U.

Read the definitions of the empty set and the universal set again. Here is the irony: The empty set is actually "universal" because it is the same in any situation; on the other hand, the universal set is not so "universal" because it depends on the problem under consideration.

The *membership notation, $x \in A$,* signifies that x is an element of set A. We can also say "x is in A" or "x belongs to A".

Remark. *Set vs. Set*

A mathematical set contains DISTINCT elements. In real life we use the term set in a more relaxed manner. Here is an example: Suppose you put two apples and three oranges in a basket. You now have a set that contains {apple, apple, orange, orange, orange}. This is a real-life set. The presentation is, however, not a mathematical set because the way it looks, there are only two distinct elements: apple and orange, and so the set must be rewritten as {apple, orange}. But this of course loses the important information that there are five objects in the basket. The remedy is to label each fruit because they are in fact all distinct (no two apples are identical), the basket can then be presented as a mathematical set: {apple 1, apple 2, orange 1, orange 2, orange 3}. Alternatively, we can also give each fruit a unique name and express the set as {Anna, Amy, Orion, October, Obama}.

Mathematics borrows a lot of terms from everyday language. We will encounter some more later. It is important to remember what the mathematical definition of a term is when we are doing mathematics. If we try to interpret a mathematical term using the meaning of the term in everyday language, we can get really confused and make some really unpleasant mistakes ◄

= EXERCISES 2.1 =

1. A set is a collection of _____ objects. The objects are called the _____ or _____ of the set.

2. Is it possible for a set to contain no elements at all?

3. Write the set $A = \{x \mid x$ is an integer between 1 and 50 that is divisible by 7$\}$ in roster form.

4. If $7 \in B$, then 7 is an _____ of B.

5. Is each of the following statements mathematically correct?

    (a) The set $\{x, y, z, u, v, w, x\}$ contains seven elements.

    (b) The two sets $\{x, y, z\}$ and $\{z, x, y\}$ are equal.

(c) The universal set, as its name suggests, contains everything in the universe. For example, the number 5 is an element in the universal set, and so is my half-brother Bingo.

(d) The empty set can be written as {Ø}.

2.2 SET OPERATIONS

There are three basic set operations.

The notation $A \cup B$ represents the *union of sets A and B*. The union contains all elements from both sets, much like when we put the contents of two baskets into one big basket. But remember we are dealing with mathematical sets so multiple identical elements will "merge" into one single element.

Example 2.2.1. *Given $A = \{1,2,3,4,5\}$ and $B = \{2,4,6,8\}$. The union of the two sets is*

$$A \cup B = \{1,2,3,4,5,6,8\}$$

Notice that elements 2 and 4 belong to both A and B, but each is listed only once in the union of the two sets◄

The notation $A \cap B$ represents the *intersection of sets A and B*. The intersection contains all elements common to both sets.

Example 2.2.2. *Given $A = \{1,2,3,4,5\}$ and $B = \{2,4,6,8\}$. The intersection of the two sets is*

$$A \cap B = \{2,4\} \blacktriangleleft$$

The notation A' represents the *complement of set A*. The complement contains all elements that do not belong to A. This operation can be performed only when the universal set U is known.

Example 2.2.3. *Given $U = \{1,2,3,4,5,6,7,8\}$ and $A = \{1,2,3,4,5\}$. The complement of set A is*

$$A' = \{6,7,8\} \blacktriangleleft$$

When two or more operations are involved in one expression, the complement precedes the union and the intersection, and operations enclosed in parentheses precede the ones outside the parentheses. For example, in the expression $A \cup B'$, we should find B' first, then find the union of A and B'; in the expression $A \cap (B \cup C)$, we should find $(B \cup C)$ first, then find the intersection of

A and (*B* ∪ *C*); in the expression *A* ∩ *B* ∪ *C* we simply perform the operations from left to right, *A* ∩ *B* first, then the union of *A* ∩ *B* and C.

Example 2.2.4. *Given U = {s, t, u, v, w, x, y, z}, A = {u, y}, B = {s, t, u, v, w}, and C = {x, y, z}. Find*

(a) *B* ∩ *A*′.

(b) *A* ∪ *B* ∩ *C*.

(c) *A* ∪ (*B* ∩ *C*).

Solution.

(a) *We perform A′ first, then take the intersection of B and A′:*

$$B \cap A' = \{s, t, u, v, w\} \cap \{s, t, v, w, x, z\} = \{s, t, v, w\}$$

(b) *We move from left to right:*

$$A \cup B \cap C = \{u, y\} \cup \{s, t, u, v, w\} \cap \{x, y, z\} = \{s, t, u, v, w, y\} \cap \{x, y, z\} = \{y\}$$

(c) *We perform the operation inside the parentheses first:*

$$A \cup (B \cap C) = \{u, y\} \cup (\{s, t, u, v, w\} \cap \{x, y, z\}) = \{u, y\} \cup \emptyset = \{u, y\} \blacktriangleleft$$

Example 2.2.5. *The names, genders, and ages of 10 children are given below:*

NAME	Ali	Bre	Cory	David	Elvis	Frank	Gina	Helen	Irene	John
GENDER	Girl	Girl	Boy	Boy	Boy	Boy	Girl	Girl	Girl	Boy
AGE	3	6	5	7	2	4.5	3.3	6	7	2.2

Find the following sets and write the answers in roster form:

(a) *G* = {all girls}.

(b) *B* = {all boys}.

(c) *Y* = {children 5 years or younger}.

(d) *R* = {children older than 5 years}.

(e) *G* ∪ *Y*.

(f) *B* ∩ *R*.

Solution.

(a) *G* = {Ali, Bre, Gina, Helen, Irene}.

(b) *B* = {Cory, David, Elvis, Frank, John}.

(c) *Y* = {Ali, Cory, Elvis, Frank, Gina, John}.

(d) *R* = {Bre, David, Helen, Irene}.

(e) $G \cup Y = \{Ali, Bre, Cory, Elvis, Frank, Gina, Helen, Irene, John\}$.

(f) $B \cap R = \{David\}$ ◄

2.3 VENN DIAGRAMS

Venn diagrams offer us a way to visualize the set operations. A rectangle is used to represent the universal set, a circle inside the rectangle is used to represent a set, and the result of a set operation is shaded in the diagram. The following are some basic Venn diagrams:

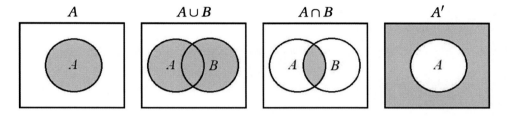

Example 2.3.1. *Shade the area in the Venn diagram that represents the set $A \cap B'$.*

Solution. *(a) First we shade the area B' which is everything outside of B. (b) We then shade A. (c) $A \cap B'$ is the area that's shaded twice.*

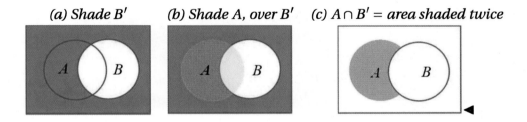

For mathematics to be useful in the real world, it has to be connected to our everyday language. The precise interpretation of language in the context of mathematics is therefore of paramount importance. Listed below are some keywords and their corresponding set operations.

1. AT LEAST ONE, or OR, indicates the union operation. Example: If A is the set of all first year students, and B is the set of all female students, then a student that belongs to $A \cup B$ is either a first year student OR a female student. Another example: If A is the set of students who have passed Math 101, and B is the set of all students who have passed English 101, then $A \cup B$ is the set that contains all students who have passed AT LEAST ONE of the two courses.

2. BOTH, or AND, indicates the intersection operation. Example: If A is the set of all first year students, and B is the set of all female students, then a student that belongs to $A \cap B$ must

be a first year AND a female student. Another example: If A is the set of students who have passed Math 101, and B is the set of all students who have passed English 101, then $A \cap B$ is the set that contains all students who have passed BOTH courses.

3. NOT, or an equivalent negation word, indicates the complement operation. Example: If A is the set of students who have passed Math 101, then A' is the set that contains all students who have NOT passed Math 101.

Remark. *OR vs. OR*

In everyday language, "A OR B" normally means either A or B, but not both, because the underlying assumption is that we are choosing only one of the two, and the fact that A and B are usually single objects reinforces this perception. In set theory, "A OR B" is defined as the union of set A and set B. Since A and B are sets, they may share some common elements, and this leads to some people defining the word OR in set theory as "A OR B OR both" just to be totally clear, and calling this the inclusive OR as opposed to the everyday OR that is called the exclusive OR.

Our next example shows that there is no contradiction between our everyday OR and the set theory OR. So it makes much more sense for us to upgrade our understanding of the word OR to this more encompassing meaning of UNION of sets. When we ask the question "do you want A OR B?" we are asking someone to pick an element from the set $A \cup B$. Whether A and B have any common elements is totally irrelevant◄

Example 2.3.2. *Use the same ten children and the same sets in example 2.2.5. If we say "anyone who is a girl OR not older than 5 years please come forward to receive a piece of chocolate from Mr. Gump", then Ali, Bre, Cory, Elvis, Frank, Gina, Helen, Irene, and John would come forward. If we take a closer look, we find that the children are exactly the set $G \cup Y$—the UNION of the set of girls and the set of children 5 years or younger. Notice here we use the connective OR in a "everyday" manner and there is no confusion whatsoever. This demonstrates the point we have just made: There is no such a thing as an "exclusive OR". All OR's are inclusive, and they mean the union of two sets◄*

Example 2.3.3. *Use the same ten children and the same sets in example 2.2.5. Describe the following sets in English:*

(a) B'

(b) $G \cap R$

(c) $B \cap Y$

(d) $B \cup Y$

Solution.

(a) $B' = \{$Children who are not boys$\}$

(b) $G \cap R = \{$Girls who are older than 5 years$\}$

(c) $B \cap Y = \{$Boys who are 5 years or younger$\}$

(d) $B \cup Y = $ {*All children who are boys, OR 5 years or younger*} ◀

= EXERCISES 2.2 & 2.3 =

1. Given $U = \{1, 2, 3, 4, 5, 6, 7, 8, 9, 10\}$, $A = \{1, 3, 5, 7, 9\}$, $B = \{2, 4, 6, 8, 10\}$, $C = \{2, 3, 5, 7\}$ and $D = \{6, 7, 8, 9, 10\}$. Find

   (a) $A \cup B$.

   (b) $A \cap B$.

   (c) A'.

   (d) $A \cap C$.

   (e) $A \cup (D' \cap C)$.

   (f) $A \cup D' \cap C$.

2. Given $U = \{\bigstar, \spadesuit, \clubsuit, \heartsuit, \diamondsuit\}$, $A = \{\spadesuit, \diamondsuit\}$, $B = \{\clubsuit, \heartsuit\}$, and $C = \{\bigstar, \heartsuit\}$. Find

   (a) $A \cup B \cup C$.

   (b) $A \cap B \cap C$.

   (c) $(B \cup C)'$.

   (d) $B' \cap C'$.

   (e) $(B \cap C)'$.

   (f) $B' \cup C'$.

   (g) $(A' \cap B)'$.

3. The names, genders, and ages of ten people are given in the table below.

NAME	Ali	Bre	Cory	David	Elvis	Frank	Gina	Helen	Irene	John
GENDER	F	F	M	M	M	M	F	F	F	M
AGE	3	5	12	21	32	45	48	60	66	72

   (a) Find the set $A = $ {people younger than 10}.

   (b) Find the set $B = $ {people older than 65}.

   (c) Find the set of people who are younger than 10 OR older than 65.

   (d) Find the set $A \cup B$.

   (e) Compare the last two results and make a connection between the connective word OR and a set operation.

   (f) Find the set $C = $ {people younger than 40}.

   (g) Find the set $D = $ {people older than 20}.

   (h) Find the set of all people who are older than 20 AND younger than 40.

   (i) Find the set $C \cap D$.

   (j) Compare the last two results and make a connection between the connective word AND and a set operation.

   (k) Find the set $F = $ {all females}.

(l) Find the set $(A \cup B) \cap F$.

4. Mathematics is always exact, but language isn't. However, mathematics cannot be fully expressed or taught or learned without the use of some language. This is a much deeper problem than most people realize. I am tired of using language... but unfortunately, I have to express this thought by using some language—I am using language to criticize language—which, for the lack of a better word, seems to be a little hypocritical.

5. Match each of the following expressions with one of the Venn diagrams.

(A) $S \cup T$ iii

(B) $S \cap T$ iv

(C) $S' \cap T'$ v

(D) $S' \cup T'$ vi

(E) $S \cap T'$ ii

(F) $S' \cap T$ i

(G) $(S \cup T)'$ v

(H) $(S \cap T)'$ vi

(I) $R \cup S \cap T$ ix

(J) $R \cup (S \cap T)$ vii

(K) $(R \cup S \cup T)'$ viii

(L) $R' \cap S' \cap T'$ viii

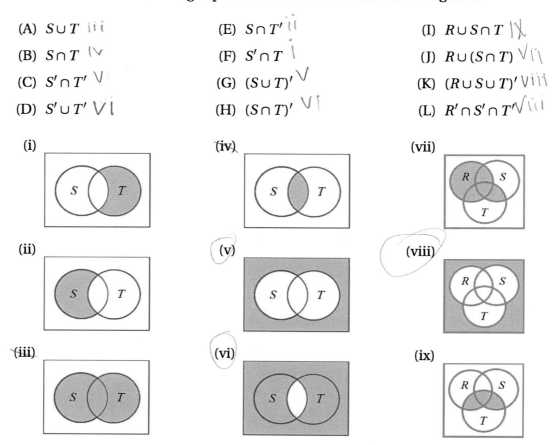

2.4 BASIC FACTS ABOUT SETS

If we take some elements from a set to form a new set, the new set is said to be a *subset* of the first set. The word "some" includes "nil" and "all". If set B is a subset of set A, we write $B \subseteq A$. For example, $\{x, y\} \subseteq \{x, y, z\}$.

Example 2.4.1. *Given the set $\{x, y, z\}$, find all its subsets.*

Solution. *We will list the subsets based on the number of elements each of them contains.*

Subsets that contain 0 elements: Ø
Subsets that contain 1 element: $\{x\}$, $\{y\}$, $\{z\}$
Subsets that contain 2 elements: $\{x, y\}$, $\{x, z\}$, $\{y, z\}$
Subsets that contain 3 elements: $\{x, y, z\}$

There are 8 subsets. We will revisit this number, 8, later. Remind me if I forget◄

Let A, B, and C represent arbitrary sets. The following are some basic facts. Most of them are self-evident. You are strongly recommended to stare at them until they make sense to you.

(i) $\varnothing \subseteq A$
(ii) $A \subseteq A$
(iii) $A \subseteq U$ for any given problem
(iv) $A \subseteq A \cup B$
(v) $B \subseteq A \cup B$
(vi) $A \cap B \subseteq A$
(vii) $A \cap B \subseteq B$
(viii) $A \cup A' = U$
(ix) $A \cap A' = \varnothing$
(x) If $C \subseteq B$ and $B \subseteq A$ then $C \subseteq A$
(xi) If $A \subseteq B$ and $B \subseteq A$ then $A = B$
(xii) $A \cup A = A$
(xiii) $A \cap A = A$
(xiv) $(A')' = A$
(xv) $(A \cup B)' = A' \cap B'$
(xvi) $(A \cap B)' = A' \cup B'$

All of the facts above should be fairly obvious except for the last two. The last two are called the *De Morgan's laws*. They are actually one formula once we accept that $(A')' = A$ is true. The example below provides one proof of the De Morgan's laws.

Example 2.4.2. *Prove* $(A \cup B)' = A' \cap B'$.

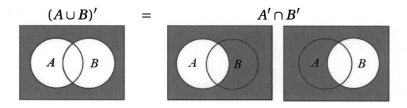

Proof. *If* $x \in (A \cup B)'$, *then* $x \notin A$ *AND* $x \notin B$ *otherwise* x *would have been in* $A \cup B$, *which implies* $x \in A' \cap B'$. *So* $(A \cup B)' \subseteq A' \cap B'$.

If $x \in A' \cap B'$, then $x \notin A$ AND $x \notin B$, which means $x \notin A \cup B$, i.e., $x \in (A \cup B)'$. So $A' \cap B' \subseteq (A \cup B)'$.

Since $(A \cup B)' \subseteq A' \cap B'$ and $A' \cap B' \subseteq (A \cup B)'$, it follows that $(A \cup B)' = A' \cap B'$ ◄

Set operations provide the foundation for logic. Human languages can be cleverly manipulated to mislead or confuse people. The following example is a story from ancient China.

Example 2.4.3. *(A White Horse is Not a Horse) The king gives an order that no horses are allowed to be taken out of his kingdom. One day a man riding a white horse tries to cross the border.*

Border guard: You can go, but not the horse.
Man: Why?
BG: The king says no horses are allowed to cross the border.
M: But this is a white horse, not a horse.
BG: How's a white horse not a horse?
M: If a white horse is a horse, then a red horse is also a horse, right?
BG: Of course.
M: So if you ask me for a white horse, then you are asking me for a horse. But since a red horse is also a horse, I can then give you a red horse. But that will not be what you ask for, therefore a white horse is not a horse.

A HORSE   NOT A HORSE?

The border guard is really confused and lets the man and his horse cross the border.

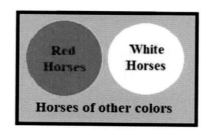

Let $A = \{$all horses$\}$, $B = \{$white horses$\}$. We see $B \subseteq A$. When the man says "if you ask me for a white horse, then you are asking me for a horse" he is switching from a subset to a set, he is saying if $B \subseteq A$ then $B = A$, which is not true in general. Stated in everyday language, it is not always easy to point out the fallacy even when you feel something just isn't right (that's how most shouting matches get started by the way).

A white horse is an element in the White Horses circle. When a white horse is requested, an element from anywhere outside the White Horses circle will not satisfy the request. The man in the story is playing with words and that can frustrate people a lot of the time◄

$$\smile \sim \smile$$
= EXERCISES 2.4 =

1. Determine whether each of the following statements is true or false.

   (a) If $x \in A$ and $A \subseteq B$, then $x \in B$. *yes*

   (b) If $A \subseteq B$, then $B' \subseteq A'$. *yes*

   (c) $A \cap B \subseteq A$. *yes*

   (d) $A \cap B' \subseteq A$. *yes*

   (e) The intersection of any two sets is a subset of either set. *yes*

   (f) A set is always a subset of the union of itself and any set. *yes*

   (g) For any two sets A and B, $(A \cap B)' = A' \cap B'$.

   (h) For any two sets A and B, $(A \cup B)' = A' \cup B'$.

   (i) For any two sets A and B, $(A \cup B)' = A' \cap B'$.

   (j) For any two sets A and B, $(A \cap B)' = A' \cup B'$.

   (k) $(A \cap B') \cup (A' \cap B) \cup (A \cap B) = A \cup B$. (Hint: Draw a Venn diagram.)

2. Use the fact that $(A \cap B)' = A' \cup B'$ and $(A')' = A$ to prove that $(A \cup B)' = A' \cap B'$.

3. Prove the identity $(A \cup B \cup C)' = A' \cap B' \cap C'$.

4. Prove the identity $(A \cap B \cap C)' = A' \cup B' \cup C'$.

5. Point out the fallacy in the argument about a white horse is not a horse in example 2.4.3.

2.5 COUNTING THE NUMBER OF ELEMENTS

Where are the numbers? Well, we can count the number of elements in a set. If set A contains 5 elements, we can express this fact by the notation $n(A) = 5$. Although this seems almost laughably simple, we have to remember that a set does not contain identical elements, and so when set operations are involved, counting is no longer as straightforward as it appears.

Example 2.5.1. *A teacher asked her class how many people owned cats, 15 raised their hands. She then asked how many owned dogs, 12 raised their hands. Finally she asked how many owned neither, and 3 raised their hands. Suppose there were 25 people in the class, how many people owned cats or dogs?*

Solution. 15 *people owned cats,* 12 *people owned dogs,* 15 + 12 = 27, *so* 27 *people owned cats or dogs, right? It looks OK, until you remember there were only* 25 *people in the class so how could there be* 27 *people who owned cats or dogs?*

There is nothing wrong with the numbers, because some cat owners were also dog owners, they raised their hands twice! They were in the intersection of the sets {cat owners} and {dog owners}.

Three people owned neither, so 25–3 = 22 *people owned cats OR dogs.* 15 *people owned cats,* 12 *people owned dogs,* 15 + 12 = 27. 27–22 = 5, *i.e.,* 5 *people were counted twice, and they were the ones who owned both cats and dogs. See the Venn diagram below.*

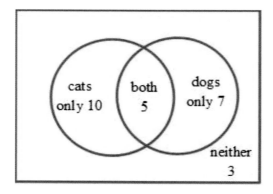

From the diagram we can see that

$$n(\text{cats}) + n(\text{dogs}) - n(\text{cats AND dogs}) = 15 + 12 - 5 = 22$$

and also

$$n(\text{cats OR dogs}) = 10 + 5 + 7 = 22$$

The two equal. This is called the inclusion-exclusion formula◄

Inclusion-exclusion formula for two sets

$$n(A \cup B) = n(A) + n(B) - n(A \cap B) \qquad (2.5.1)$$

Example 2.5.2. *Let's revisit the previous example and see how we can use the inclusion-exclusion formula to solve the problem. Let C be the set of people who owned cats, and D be the set of people who owned dogs. It is clear that* $n(C) = 15$ *and* $n(D) = 12$. *Since there were* 3 *people who owned neither cats nor dogs, and there were* 25 *people in the class, that means* 22 *people owned either* *cats OR dogs, i.e.,* $n(C \cup D) = 22$. *According to (2.5.1)*

$$n(C \cup D) = n(C) + n(D) - n(C \cap D)$$
$$22 = 15 + 12 - n(C \cap D)$$

so

$$n(C \cap D) = 5$$

i.e., 5 *people owned both cats AND dogs*◄

We now progress to three sets. Here is a typical three-set Venn diagram:

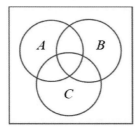

As an exercise, please verify that $(A \cup B) \cap C = (A \cap C) \cup (B \cap C)$.

We introduce the three-set Venn diagram to show that the inclusion-exclusion formula can be extended to more than two sets. The inclusion-exclusion formula quickly becomes too overwhelming when the number of sets increases, though.

> **Inclusion-exclusion formula for three sets**
>
> $$n(A \cup B \cup C) = n(A) + n(B) + n(C) - n(A \cap B) - n(A \cap C) - n(B \cap C) + n(A \cap B \cap C)$$
> (2.5.2)

The proof of the formula above is not difficult. The key step is to write $n(A \cup B \cup C) = n((A \cup B) \cap C)$ and then use the inclusion-exclusion formula for two sets (2.5.1).

Some problems that involve three sets can be solved by a Venn diagram alone without relying on the inclusion-exclusion formula. Here is an example.

Example 2.5.3. *A survey of 48 people revealed the following:* 21 *people owned cats,* 23 *owned dogs,* 14 *owned pigs,* 10 *owned both cats and dogs,* 5 *owned both cats and pigs,* 8 *owned both dogs and pigs, and* 3 *owned all three animals.*

   (a) How many owned none of the three animals?

    (b) How many owned only pigs?

    (c) How many owned cats or dogs but not pigs?

    (d) How many owned cats and dogs but not pigs?

Solution.

1. We use three circles to represent cat owners, dog owners, and pig owners, and draw them inside a rectangle that represents the 48 people surveyed

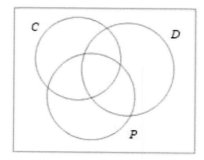

2. Now put the number 3, which is the number of people who owned all three animals, in the center region that is the intersection of the three circles

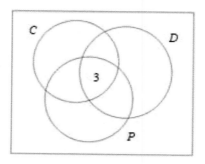

3. Reading up the list of the numbers given, we see that 8 owned dogs and pigs, so the intersection of the D circle and the P circle contains 8 people, but 3 people are already there, that leaves 5 people to be added to that region

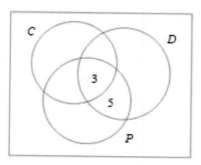

4. *Make sure you see how the next two numbers are calculated when we count the number of people who owned both cats and pigs, and the number of people who owned both cats and dogs*

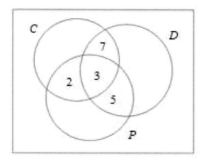

5. *Moving up the list of numbers given we see that 14 people owned pigs, so the P circle contains 14. But there are 2 + 3 + 5 = 10 people already there, that leaves 4 people to be added to that region*

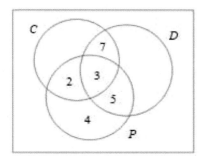

6. *I will leave it to you to verify the next two numbers in the C and P circles*

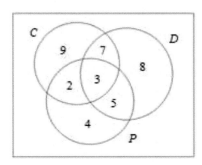

7. *Finally you add all the numbers in the diagram together: 9 + 7 + 8 + 2 + 3 + 5 + 4 = 38, then subtract that from the total number of people 48: 48–38 = 10, and put this number outside the three circles, but inside the rectangle. This is the number of people who owned none of the three animals*

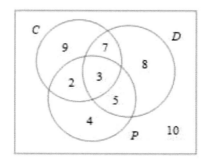

The Venn diagram is now complete, and we are ready to answer the questions:

(a) 10, *the number outside the three circles.*

(b) 4, *the number in the P circle, but outside the C and D circles.*

(c) $9 + 7 + 8 = 24$, *the sum of the numbers inside the C circle OR the D circle but outside the P circle.*

(d) 7, *the number inside the C circle AND the D circle, but outside the P circle*◄

The last example shows how we may solve a three-set problem using a Venn diagram by starting from the center—the intersection of all three sets.

Remark. *When there are four or more sets involved, Venn diagrams (with circles) can no longer be used to illustrate the situation. People who are intrigued by this fact can do some research on their own. The inclusion-exclusion formula, however, can be extended to four or more sets*◄

= EXERCISES 2.5 =

1. Just a reminder here, the connective word OR is always inclusive, and it tells us to perform the set operation UNION. The connective word AND tells us to perform the set operation INTERSECTION. Now repeat after me ten times.

2. There are six cars in a parking lot: a red Chevrolet Corvette, a red Chevrolet Camaro, a blue Chevrolet Cavalier, a red Ford F-150, a black Ford Mustang, and a black Dodge Neon.

    (a) Find the set C of all Chevrolet cars.

    (b) Find the set R of all red cars.

    (c) Find the set $C \cup R$.

    (d) Find the set of all the cars that are Chevrolet OR red. How does the answer compare to the answer of the previous question?

    (e) Find the set $C \cap R$.

    (f) Find the set of all red Chevrolet cars. How does the answer compare to the answer of the previous question?

    (g) Find $n(C)$, $n(R)$, $n(C \cup R)$, and $n(C \cap R)$.

    (h) Verify that $n(C \cup R) = n(C) + n(R) - n(C \cap R)$.

3. Last year the College of Peaceful Seas admitted 800 freshmen. All of them had to take the math and English placement exams. After the exams it was determined that 500 of the freshmen needed to take the math refresher course, 340 needed to take the English refresher course, but 220 of them didn't need to take either. How many freshmen needed to take both the math refresher course and the English refresher course?

4. Last year the College of Rugged Mountains admitted 79 freshmen. All of them had to take the math and English placement exams. After the exams it was determined that 50 of the freshmen needed to take either the math refresher course or the English refresher course, 35 needed to take the math refresher course and 9 needed to take both the math refresher course and the English refresher course.

    (a) How many freshmen needed to take neither the math nor the English refresher course?

    (b) How many freshmen needed to take the English refresher course?

    (c) How many freshmen needed to take ONLY the math refresher course?

    (d) How many freshmen needed to take only the English refresher course?

5. Once thirteen bad guys tried to overpower Bruce Lee, but failed miserably, of course.

Eight of them got hit by powerful straight punches, ten of them got hit by even more powerful side thrust kicks, two of them were smart enough to play dead and didn't get hit at all. How many of the bad guys got hit by both straight punches and side thrust kicks?

6. A standardized test consists of three parts. 6000 people took the test last year. 3400 people passed part I, 3400 people passed part II, 3900 people passed part III, 2600 people passed parts I and II, 2700 people passed parts I and III, 3000 people passed parts II and III, and 2500 people passed all three parts.

    (a) Draw a three-circle Venn diagram based on the given information.

    (b) How many people passed none of the three parts of the test?

    (c) How many people passed parts I or II, but not III?

(d) How many people passed only part III?

(e) How many people passed at least one part?

(f) How many people passed exactly one part?

(g) How many people passed at least two parts?

(h) How many people passed exactly two parts?

7. A long time ago in a communist country far, far away people were required by law to wear at least one of three colors: red, green, and blue. Anyone not wearing one of these three colors was considered an anti-revolution bad guy and could be sent to a reform farm forever. A random sampling of 87 people revealed that 5 people could be considered anti-revolution bad guys, 25 wore red, 55 wore green, 19 people wore red and green, 5 people wore red and blue, 22 people wore green and blue, and 3 people wore red, green and blue.

(a) Draw a three-circle Venn diagram based on the given information.

(b) How many people wore only blue?

(c) How many people wore only red?

(d) How many people wore only green?

Chapter 3

Linear Algebra

"Linear" means "like straight line". In math, a two-variable linear equation takes the form $Ax + By = C$ where A, B, and C are constants with $AB \neq 0$. The graph of this equation is a straight line. If either x or y is raised to a power other than 1, e.g., $x^2 + y = 5$, the graph of the equation is no longer a straight line. From this fact, the term "linear" is extended to describe all equations or functions that involve only variables that are raised to the power 1, with each term consists of only one variable. For example, $2x + 3y - 4z = 5$ is a three-variable linear equation. The graph of this equation is a flat plane in the three-dimensional space, not a straight line. The equation $2xy - 3z = 4$ is not a linear equation because the term $2xy$ consists of two variables.

3.1 THE GAUSS-JORDAN ELIMINATION METHOD

A two-variable-two-equation linear system (2 by 2 system) such as

$$\begin{cases} x + 2y = 5 \\ 3x - y = -13 \end{cases}$$

can be solved by the substitution method. We first solve the first equation for x: $x = 5 - 2y$, then substitute $5 - 2y$ for x in the second equation and the second equation becomes $3(5-2y) - y = -13$. After solving this last equation for y we get $y = 4$, it then follows that $x = -3$. The system is now solved, and we may write the answer as a pair of numbers in parentheses: $(-3, 4)$.

The substitution method is convenient for solving a small 2 by 2 system. It becomes inefficient when we need to solve larger systems. The following is a 4 by 4 system and you can imagine how much work would be involved if we try to solve it using the substitution method.

$$\begin{cases} w + x + 2y - z = 2 \\ -2w + 4x - y + 3z = 6 \\ w + 3x + 3y + 4z = 19 \\ -3w + 2x + y - 5z = -20 \end{cases}$$

Enter the *Gauss-Jordan elimination method.* (The name is a little misleading. Gauss, a great mathematician, wasn't directly involved in the invention of the method. It was actually more closely associated with Sir Isaac Newton. The method was also recorded in ancient China.) If you are proficient in a programming language, the method can be coded in a few lines. If you don't do programming at all, you can still learn a lot from the method. You learn how a seemingly formidable task

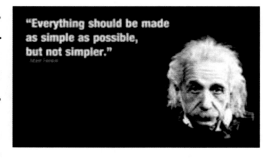

such as solving a 100 by 100 system can be accomplished by a method that employs THREE basic operations! You also learn the importance of following rules when we know the rules will prevail—I know this sounds silly, but still, sometimes people forget the basics.

There are three facts about a system of equations:

1. Interchanging the positions of any two equations in a system does not change the solution to the system. For example, the following two systems

$$\begin{cases} x + 2y = 5 \\ 3x - y = -13 \end{cases} \quad \text{and} \quad \begin{cases} 3x - y = -13 \\ x + 2y = 5 \end{cases}$$

   have the exact same solution $(-3, 4)$ because they consist of the same two equations in different order.

2. The solution to an equation remains the same if we multiply both sides of the equation by a non-zero number. For example, the following two equations

$$2x + 1 = 6 \quad \text{and} \quad 6x + 3 = 18$$

   have the exact same solution $x = 5/2$ because the second equation is simply the first one multiplied by 3.

3. Replacing one equation in a system with the sum of this equation and another equation in the system does not change the solution to the system. For example, the following two systems

$$\begin{cases} x + 2y = 5 \\ 3x - y = -13 \end{cases} \quad \text{and} \quad \begin{cases} x + 2y = 5 \\ 4x + y = -8 \end{cases}$$

have the exact same solution $(-3, 4)$ because the second system is the first system with its second equation replaced by the sum of its first two equations.

To perform the Gauss-Jordan elimination method, we first write a system as a *matrix*. A matrix is a rectangular array of numbers, enclosed in a pair of brackets. The system

$$\begin{cases} x + 2y = 5 \\ 3x - y = -13 \end{cases}$$

can be represented by the following matrix

$$\left[\begin{array}{cc|c} 1 & 2 & 5 \\ 3 & -1 & -13 \end{array} \right]$$

with the understanding that the numbers in the first column are the coefficients of the variable x in all the equations, the numbers in the second column are the coefficients of the variable y in all the equations, the numbers in the last column are the constants on the right-hand side of the equal signs, and each row of numbers corresponds to one equation in the system. The vertical line separating the last column is optional as long as we understand what each column represents.

Example 3.1.1. *Convert the following matrices to corresponding systems of equations. Use x, y, z as variables.*

(a) $\left[\begin{array}{ccc|c} 2 & 0 & 3 & 9 \\ 1 & 5 & -7 & 8 \\ 4 & -1 & 6 & -10 \end{array} \right]$
  (b) $\left[\begin{array}{ccc|c} 1 & 0 & 0 & 6 \\ 0 & 1 & 0 & -5 \\ 0 & 0 & 1 & 3 \end{array} \right]$

Solution.

(a) $\begin{cases} 2x \quad\quad + 3z = 9 \\ x + 5y - 7z = 8 \\ 4x - y + 6z = -10 \end{cases}$
  (b) $\begin{cases} x \quad\quad\quad = 6 \\ \quad y \quad\quad = -5 \\ \quad\quad z = 3 \end{cases}$

The second matrix is special, it gives the values of x, y, and z in the last column: $(6, -5, 3)$ ◄

When we express a system in the form of a matrix, the three facts about a system of equations can be translated into the following three *elementary matrix row operations*:

Elementary matrix row operations
1. Interchanging two rows.
2. Multiplying a row by a non-zero number.
3. Adding the multiple of one row to another row.

We have all the tools and regulations in place now. Let's go do some conquering first, then explain and sort out all the details after that.

Example 3.1.2. *Use the Gauss-Jordan elimination method (I know we haven't officially delineated the method, yet. But humans have watched tons of things they have never seen before, and survived, haven't they?) to solve the system*

$$\begin{cases} 2w & - & x & + & 3y & - & z & = & 5 \\ w & + & x & - & y & + & z & = & 4 \\ 3w & - & x & - & y & - & z & = & -6 \\ 4w & - & 2x & - & 5y & + & 3z & = & -3 \end{cases}$$

Solution. *First we write the system in the matrix form*

$$\left[\begin{array}{cccc|c} 2 & -1 & 3 & -1 & 5 \\ 1 & 1 & -1 & 1 & 4 \\ 3 & -1 & -1 & -1 & -6 \\ 4 & -2 & -5 & 3 & -3 \end{array} \right]$$

*We can then perform the three row operations on this matrix because the operations will not alter the solution to the system. We need to have a goal in mind though, because if we simply perform the three row operations randomly, we probably will never find the **Solution**. Inspired by the previous example, we want our final matrix to look like*

$$\left[\begin{array}{cccc|c} 1 & 0 & 0 & 0 & * \\ 0 & 1 & 0 & 0 & * \\ 0 & 0 & 1 & 0 & * \\ 0 & 0 & 0 & 1 & * \end{array} \right]$$

*where each asterisk stands for a number. We will do the transformation systematically, **one column at a time—this is very important**. If you deviate from this route, you may get into unnecessary trouble.*

Before we proceed, we need to introduce a few bookkeeping notations:

(a) R1 stands for the first row in the matrix, R2 the second row, etc.

(b) R1 ↔ R3 indicates that the first row and the third row exchange their positions.

(c) 5R3 means every number in the third row is multiplied by 5.

(d) 5R3 + R4 tells us to add five times of the third row to the fourth row (note here only the fourth row changes, the third row remains the same).

Have your pencil and scratch paper ready. Verify every step of the following process.

Start with the initial matrix

$$\left[\begin{array}{cccc|c} 2 & -1 & 3 & -1 & 5 \\ 1 & 1 & -1 & 1 & 4 \\ 3 & -1 & -1 & -1 & -6 \\ 4 & -2 & -5 & 3 & -3 \end{array}\right]$$

We want the first number in the first column to be a 1, we can achieve this by interchanging row 1 and row 2

$$\xrightarrow{R1 \leftrightarrow R2} \left[\begin{array}{cccc|c} 1 & 1 & -1 & 1 & 4 \\ 2 & -1 & 3 & -1 & 5 \\ 3 & -1 & -1 & -1 & -6 \\ 4 & -2 & -5 & 3 & -3 \end{array}\right]$$

Now convert the other elements of column 1 to zeros by adding appropriate multiples of row 1 to each row below. For example, the operation $-2R1 + R2$ tells us that row 2 is being modified by adding -2 times of row 1 to it. Note that row 1 stays the same

$-2R1:$	-2	-2	2	-2	-8
$+)\ \ R2:$	2	-1	3	-1	5
$New\ R2:$	0	-3	5	-3	-3

$$\begin{array}{c} -2R1 + R2 \\ -3R1 + R3 \\ -4R1 + R4 \\ \xrightarrow{\hspace{2cm}} \end{array} \left[\begin{array}{cccc|c} 1 & 1 & -1 & 1 & 4 \\ 0 & -3 & 5 & -3 & -3 \\ 0 & -4 & 2 & -4 & -18 \\ 0 & -6 & -1 & -1 & -19 \end{array}\right]$$

Next we convert the second number in the second column to a 1

$$\xrightarrow{-\frac{1}{3}R2} \left[\begin{array}{cccc|c} 1 & 1 & -1 & 1 & 4 \\ 0 & 1 & -5/3 & 1 & 1 \\ 0 & -4 & 2 & -4 & -18 \\ 0 & -6 & -1 & -1 & -19 \end{array}\right]$$

Then eliminate the other numbers in the second column

$$\begin{array}{c} -R2 + R1 \\ 4R2 + R3 \\ 6R2 + R4 \\ \xrightarrow{\hspace{2cm}} \end{array} \left[\begin{array}{cccc|c} 1 & 0 & 2/3 & 0 & 3 \\ 0 & 1 & -5/3 & 1 & 1 \\ 0 & 0 & -14/3 & 0 & -14 \\ 0 & 0 & -11 & 5 & -13 \end{array}\right]$$

Ready for column 3 now. Turn the third number in the third column to a 1

$$\xrightarrow{-\frac{3}{14}R3} \left[\begin{array}{cccc|c} 1 & 0 & 2/3 & 0 & 3 \\ 0 & 1 & -5/3 & 1 & 1 \\ 0 & 0 & 1 & 0 & 3 \\ 0 & 0 & -11 & 5 & -13 \end{array}\right]$$

Eliminate the other numbers in column 3

$$\begin{array}{c} (-2/3)R3 + R1 \\ (5/3)R3 + R2 \\ 11R3 + R4 \\ \longrightarrow \end{array} \left[\begin{array}{cccc|c} 1 & 0 & 0 & 0 & 1 \\ 0 & 1 & 0 & 1 & 6 \\ 0 & 0 & 1 & 0 & 3 \\ 0 & 0 & 0 & 5 & 20 \end{array}\right]$$

Make the fourth number in the fourth column a 1

$$\begin{array}{c} \dfrac{1}{5}R4 \\ \longrightarrow \end{array} \left[\begin{array}{cccc|c} 1 & 0 & 0 & 0 & 1 \\ 0 & 1 & 0 & 1 & 6 \\ 0 & 0 & 1 & 0 & 3 \\ 0 & 0 & 0 & 1 & 4 \end{array}\right]$$

Eliminate the rest of column 4

$$\begin{array}{c} -R4 + R2 \\ \longrightarrow \end{array} \left[\begin{array}{cccc|c} 1 & 0 & 0 & 0 & 1 \\ 0 & 1 & 0 & 0 & 2 \\ 0 & 0 & 1 & 0 & 3 \\ 0 & 0 & 0 & 1 & 4 \end{array}\right]$$

The desired form has been achieved. The solution is $(1,2,3,4)$ ◀

The evolution of the matrix in this process is pictured below. Notice how the matrix is converted column by column.

$$\left[\begin{array}{cccc|c} * & * & * & * & * \\ * & * & * & * & * \\ * & * & * & * & * \\ * & * & * & * & * \end{array}\right]$$

$$\longrightarrow \left[\begin{array}{cccc|c} 1 & * & * & * & * \\ * & * & * & * & * \\ * & * & * & * & * \\ * & * & * & * & * \end{array}\right] \longrightarrow \left[\begin{array}{cccc|c} 1 & * & * & * & * \\ 0 & * & * & * & * \\ 0 & * & * & * & * \\ 0 & * & * & * & * \end{array}\right]$$

$$\longrightarrow \left[\begin{array}{cccc|c} 1 & * & * & * & * \\ 0 & 1 & * & * & * \\ 0 & * & * & * & * \\ 0 & * & * & * & * \end{array}\right] \longrightarrow \left[\begin{array}{cccc|c} 1 & 0 & * & * & * \\ 0 & 1 & * & * & * \\ 0 & 0 & * & * & * \\ 0 & 0 & * & * & * \end{array}\right]$$

$$\longrightarrow \left[\begin{array}{cccc|c} 1 & 0 & * & * & * \\ 0 & 1 & * & * & * \\ 0 & 0 & 1 & * & * \\ 0 & 0 & * & * & * \end{array}\right] \longrightarrow \left[\begin{array}{cccc|c} 1 & 0 & 0 & * & * \\ 0 & 1 & 0 & * & * \\ 0 & 0 & 1 & * & * \\ 0 & 0 & 0 & * & * \end{array}\right]$$

$$\longrightarrow \begin{bmatrix} 1 & 0 & 0 & * & * \\ 0 & 1 & 0 & * & * \\ 0 & 0 & 1 & * & * \\ 0 & 0 & 0 & 1 & * \end{bmatrix} \longrightarrow \begin{bmatrix} 1 & 0 & 0 & 0 & * \\ 0 & 1 & 0 & 0 & * \\ 0 & 0 & 1 & 0 & * \\ 0 & 0 & 0 & 1 & * \end{bmatrix}$$

Let's solve one more system before we move on.

Example 3.1.3. *Solve the system*

$$\begin{cases} x & - & 3y & + & 4z & = & -6 \\ 3x & + & y & + & 9z & = & -2 \\ 4x & - & y & - & z & = & 21 \end{cases}$$

Solution.

$$\begin{bmatrix} 1 & -3 & 4 & -6 \\ 3 & 1 & 9 & -2 \\ 4 & -1 & -1 & 21 \end{bmatrix}$$

$$\begin{matrix} -3R1+R2 \\ -4R1+R3 \\ \longrightarrow \end{matrix} \begin{bmatrix} 1 & -3 & 4 & -6 \\ 0 & 10 & -3 & 16 \\ 0 & 11 & -17 & 45 \end{bmatrix}$$

$$\begin{matrix} \dfrac{1}{10}R2 \\ \longrightarrow \end{matrix} \begin{bmatrix} 1 & -3 & 4 & -6 \\ 0 & 1 & -3/10 & 8/5 \\ 0 & 11 & -17 & 45 \end{bmatrix}$$

$$\begin{matrix} 3R2+R1 \\ -11R2+R3 \\ \longrightarrow \end{matrix} \begin{bmatrix} 1 & 0 & 31/10 & -6/5 \\ 0 & 1 & -3/10 & 8/5 \\ 0 & 0 & -137/10 & 137/5 \end{bmatrix}$$

$$\begin{matrix} -\dfrac{10}{137}R3 \\ \longrightarrow \end{matrix} \begin{bmatrix} 1 & 0 & 31/10 & -6/5 \\ 0 & 1 & -3/10 & 8/5 \\ 0 & 0 & 1 & -2 \end{bmatrix}$$

$$\begin{matrix} (-31/10)R3+R1 \\ (3/10)R3+R2 \\ \longrightarrow \end{matrix} \begin{bmatrix} 1 & 0 & 0 & 5 \\ 0 & 1 & 0 & 1 \\ 0 & 0 & 1 & -2 \end{bmatrix}$$

Solution: $(5, 1, -2)$ ◀

Some systems have no solutions. For example, if a system consists of the two equations: $x + y = 1$ and $x + y = 2$, then there is no solution because $x + y$ cannot possibly be 1 and 2 at the same time. If we try to solve such a system using the Gauss-Jordan method, we will produce some impossible equation and that's how we know we have a system that has no solution.

Example 3.1.4. *Solve the system*

$$\begin{cases} x & - & y & + & z & = & 2 \\ x & + & y & + & z & = & 3 \\ x & & & + & z & = & 4 \end{cases}$$

Solution.

$$\begin{bmatrix} 1 & -1 & 1 & 2 \\ 1 & 1 & 1 & 3 \\ 1 & 0 & 1 & 4 \end{bmatrix}$$

$$\begin{matrix} -R1+R2 \\ -R1+R3 \\ \xrightarrow{\hspace{1cm}} \end{matrix} \begin{bmatrix} 1 & -1 & 1 & 2 \\ 0 & 2 & 0 & 1 \\ 0 & 1 & 0 & 2 \end{bmatrix}$$

$$\begin{matrix} \frac{1}{2}R2 \\ \xrightarrow{\hspace{1cm}} \end{matrix} \begin{bmatrix} 1 & -1 & 1 & 2 \\ 0 & 1 & 0 & 1/2 \\ 0 & 1 & 0 & 2 \end{bmatrix}$$

$$\begin{matrix} R2+R1 \\ -R2+R3 \\ \xrightarrow{\hspace{1cm}} \end{matrix} \begin{bmatrix} 1 & 0 & 1 & 5/2 \\ 0 & 1 & 0 & 1/2 \\ 0 & 0 & 0 & 3/2 \end{bmatrix}$$

The system has no solution because the last row represents the equation $0 = 3/2$*, which is impossible*◀

Some systems have infinitely many solutions. For example, if a system consists of the two equations $x+2y = 5$ and $2x+4y = 10$, then it has infinitely many solutions, because the two equations are equivalent—the second equation is simply the first one multiplied by 2. Here are a few solutions: $(1,2)$, $(5,0)$, $(3,1)$, and you can find as many as you like. If we try to solve such a system using the Gauss-Jordan method, we will produce some rows that are entirely zeros.

Example 3.1.5. *Solve the system*

$$\begin{cases} x & - & 2y & + & z & = & 5 \\ x & + & y & + & 7z & = & -1 \\ 2x & - & y & + & 8z & = & 4 \end{cases}$$

Solution.

$$\begin{bmatrix} 1 & -2 & 1 & 5 \\ 1 & 1 & 7 & -1 \\ 2 & -1 & 8 & 4 \end{bmatrix}$$

$$\begin{array}{c} -R1 + R2 \\ -2R1 + R3 \\ \hline \end{array} \begin{bmatrix} 1 & -2 & 1 & 5 \\ 0 & 3 & 6 & -6 \\ 0 & 3 & 6 & -6 \end{bmatrix}$$

$$\begin{array}{c} \frac{1}{3}R2 \\ \hline \end{array} \begin{bmatrix} 1 & -2 & 1 & 5 \\ 0 & 1 & 2 & -2 \\ 0 & 3 & 6 & -6 \end{bmatrix}$$

$$\begin{array}{c} 2R2 + R1 \\ -3R2 + R3 \\ \hline \end{array} \begin{bmatrix} 1 & 0 & 5 & 1 \\ 0 & 1 & 2 & -2 \\ 0 & 0 & 0 & 0 \end{bmatrix}$$

We see the last row is all zeros, which does not provide any useful information, so we have only two equations left:

$$\begin{cases} x & + & 5z & = & 1 \\ & y & + & 2z & = & -2 \end{cases}$$

At this point we move the z terms to the right-hand sides of the equations, and note that we can choose any value as z. Every z value produces one solution to the system.

$$\begin{cases} x & = & 1 - 5z \\ y & = & -2 - 2z \\ z & = & any\ real\ number \end{cases}$$

From the last set of equations we can generate as many solutions as we like. For example, if we choose z = 0 we get one solution (1,–2,0); if we choose z = 1 we get another solution (–4,–4,1). You should substitute a few more numbers for z to obtain a few more solutions◄

≡ **EXERCISES 3.1** ≡

1. If you learn to dissect a turkey with a knife, you will have a good chance at dissecting a bull with the same knife. If you cut a turkey with an electric knife, and the same knife is not powerful enough to cut a bull, then you can't handle a bull. The moral of the story? Correct methods and understanding can carry one farther than raw power can.

2. Use the Gauss-Jordan elimination method to solve the system. If you follow the suggested

steps and do all the arithmetic correctly, there should be no fractions in the entire process.

(a) $\begin{cases} x - y - z = -2 \\ 2x - y - 3z = -7 \\ x + y - z = -4 \end{cases}$

(c) $\begin{cases} 2x - y = -1 \\ x - y = -4 \end{cases}$

(b) $\begin{cases} x + z = 7 \\ -2x + y = 5 \\ x + y + 2z = 18 \end{cases}$

(d) $\begin{cases} x + y + 2w = 19 \\ x + 2y + 3w = 28 \\ x + 3y + z + 2w = 21 \\ 2x + y + z = 4 \end{cases}$

3. Use the Gauss-Jordan elimination method to solve the following system. There will be a lot of fractions involved in the process. Don't be intimidated. Stick to the rules and push forward until you reach the final answer. For people who are really scared of fractions, using high precision decimals is acceptable (for example, approximating 1/3 with 0.3333333333 is OK, approximating 1/3 with 0.3 is not that OK). However, if you want to challenge yourself, stick with fractions all the way through.

(a) $\begin{cases} 9x - 8y + 7z = -6 \\ 5x - 4y + 3z = -2 \\ 10x + 8y - z = 12 \end{cases}$

(b) $\begin{cases} 4x + 3y + 2z = 1 \\ 5x + 7y + 6z = 8 \\ -2x - 5y + z = -9 \end{cases}$

(c) $\begin{cases} 11x + 13y = 17 \\ 5x + 7y = 2 \end{cases}$

4. Each of the following systems has infinitely many solutions. Find three solutions for each system.

(a) $\begin{cases} x - 3y = 17 \\ -2x + 6y = -34 \end{cases}$

(b) $\begin{cases} x + z = 7 \\ -2x + y - 3z = 5 \\ x - y + 2z = -12 \end{cases}$

(c) $\begin{cases} -x + 2y + z = 6 \\ 2x + y + z = 7 \\ x + 3y + 2z = 13 \end{cases}$

5. Application problems (or word problems) are harder to solve because they require setting up mathematical equations from information described in everyday language. The mathematical flavor of such problems comes from two main steps:

- Use a letter to represent an unknown, such as "let x = the number of people who attended the conference".

- Put an equal sign between two different expressions that represent a common quantity. At least one of the expressions must involve some unknowns. An equation like $2 + 3 = 5$ is correct but useless because everything in it is known so we can't use it to find an unknown, but one like $0.7x^2 = 35$ is useful if we are told that the area of a rectangle is 35 in^2 and its width is 70% of its length and we use x to represent the length of the rectangle.

The rest is reading comprehension, and a lot of practice. Here is an example: Al needs to create 50 gallons of 70% alcohol solution. He has access to two alcohol solutions, a 40% alcohol solution and a 90% alcohol solution. How many gallons of each solution should Al use to create his desired solution? To find the answer, we need to find two unknowns—the volume of the 40% solution needed and the volume of the 90% solution needed. So we let x = the volume of the 40% solution needed and y = the volume of the 90% solution needed. One equation is easier to see, which is $x + y = 50$ because Al needs to have 50 gallons of the final solution and the 50 gallons must be the sum of the volumes of the 40% solution and the 90% solution. Another equation comes from the fact that the total amount of alcohol in the final solution comes from the amount of alcohol in the 40% solution and the amount of alcohol in the 90% solution combined. The amount of alcohol in the 40% solution is $40\%x = 0.4x$, the amount of alcohol in the 90% solution is $90\%y = 0.9y$, and the amount of alcohol in the final solution is $70\%(50) = 35$. Therefore $0.4x + 0.9y = 35$. Putting the two equations together gives us a system of two equations with two unknowns

$$\begin{cases} x + y = 50 \\ 0.4x + 0.9y = 35 \end{cases}$$

Please solve this system and tell Al the answer.

6. Brothers J, K and L inherited a large sum of money from their father. J inherited twice as much as L did, and the amount K inherited is half of the sum of what J and L inherited. Suppose the total amount the three brothers inherited was $90,000,000. How much did each brother inherit? Set up a system of three equations and solve it with the Gauss-Jordan method. You might be able to find the answer by trial-and-error, but that's like adding 2 and 3 together by counting your fingers. If you insist on doing the trial-and-error method, try replacing the word "twice" with "1.35 times", "half" with "73%", and 90,000,000 with 87,777,321 and see if you can still do it. The person who goes with setting up a system and solving it will have no trouble solving the system with these new numbers. The person who goes with the trial-and-error method will have a hard time repeating the same feat.

7. A total of 380 people watched a high school football game. Some people paid $10/person to watch, some paid $6/person to watch, and some paid $0 to watch. The revenue from the ticket sales was $3120, and the number of $10 tickets sold was 120 more than the number of $6 tickets sold. Find the number of tickets sold at each price. (Again, don't use trial-and-error, that's very uncool.)

8. Two kinds of tree nuts are available for some zoo monkeys. The first kind contains 15% iron and 4% calcium, the second kind contains 7% iron and 6% calcium. Each monkey needs 10 grams of iron and 5 grams of calcium each day. How many grams of each kind of

nuts must be given to each monkey each day to meet the iron and calcium requirements?
Let x be the amount of the first kind of nuts, and y the amount of the second kind. Choose
the system that can be used to solve the problem.

(A) $\begin{cases} 15x + 7y = 1000 \\ 4x + 6y = 500 \end{cases}$

(C) $\begin{cases} .15x + .04y = 10 \\ .07x + .06y = 5 \end{cases}$

(B) $\begin{cases} .15x + .7y = 10 \\ .4x + .6y = 5 \end{cases}$

(D) $\begin{cases} 15\%x + 4\%y = 1000 \\ 7\%x + 6\%y = 500 \end{cases}$

9. Find the system that can be used to solve the following problem: The math department
spent a total of \$8720 on three types of calculators. A graphing calculator costs \$80, a
scientific calculator costs \$12, and a four-function basic calculator costs \$4. If the number
of scientific calculators was half of the number of graphing calculators, and the number
of graphing calculators was 20 more than the number of scientific and basic calculators
combined, how many of each type of calculators did the math department purchase? Let x
be the number of graphing calculators, y the number of scientific calculators, and z the
number of basic calculators.

(A) $\begin{cases} 80x + 12y + 4z = 8720 \\ x = \frac{1}{2}y \\ x + 20 = y + z \end{cases}$

(C) $\begin{cases} 80x + 12y + 4z = 8720 \\ x - 2y = 0 \\ x - y - z = 20 \end{cases}$

(B) $\begin{cases} 80x + 12y + 4z = 8720 \\ x + y = \frac{1}{2} \\ x + y + z = 20 \end{cases}$

(D) $\begin{cases} x + y + z = 8720 \\ x = 2y \\ x = y + z + 20 \end{cases}$

3.2　LINEAR PROGRAMMING—INTRODUCTION

Linear programming is fascinating, mathematically and historically. It belongs to a broader
subject called mathematical optimization. In a nutshell, it is about getting the "best deal" under
certain constraints.

Below is a typical textbook linear programming problem—a little artificial, but nonetheless
possesses the spirit of a linear programming problem.

Example 3.2.1. *Patrick is a part-time math teacher. To make ends meet, he needs to teach at two
different colleges. A class at college X pays \$3000 and requires 6 hours per week (class preparation,
grading, traveling, etc.); a class at college Y pays \$3500 and requires 9 hours per week. If Patrick
can teach at most 5 classes and can devote at most 36 hours per week to teaching, how many
classes should he teach at each college in order to maximize his total income?*

Discussion: We will find the solution later. For now let's look at the numbers to appreciate the challenge facing us. College Y pays better, so why doesn't Patrick just teach five classes at college Y? He can't do that because he has a maximum of 36 hours available per week. If he teaches five classes at Y, he will spend a total of $5 \times 9 = 45$ hours per week teaching, and that's over his available time of 36 hours.

At this point you may feel you can solve this problem by reducing the number of classes at college Y to 4 and increasing the number of classes at college X to 1, then check the number of hours required again to see if everything is OK. If not OK then you adjust the numbers again and check again. This method indeed will solve this particular problem because there are only FIVE classes between TWO colleges. But you can foresee the difficulty if there are three colleges to choose from and the number of classes is significantly increased. In real life situations, there could be hundreds or thousands of constraints that need to be satisfied simultaneously, and adjusting the numbers one at a time is impractical.

Setting up: We do not have the tools to solve this problem yet, but we can organize the given information and set up the problem in the standard form.

Let x and y be the numbers of classes Patrick intends to teach at colleges X and Y, respectively. We can express the following quantities in terms of x and y:

Total amount of income: $3000x + 3500y$
Total number of classes: $x + y$
Total amount of time required per week: $6x + 9y$

Patrick's goal is to maximize income but teach at most 5 classes and spend at most 36 hours per week. Putting everything together we can express this problem in the following form:

$$\text{Maximize} \quad 3000x + 3500y$$
$$\text{Subject to} \quad \begin{cases} x + y \le 5 \\ 6x + 9y \le 36 \\ x \ge 0 \\ y \ge 0 \end{cases}$$

The function $3000x + 3500y$ is called the objective function. The set of inequalities is called the constraints. The first inequality is the constraint on the number of classes, the second is on the number of hours. The last two are not explicitly expressed in the problem, but are obvious and should be included because the number of classes cannot be negative◄

All linear programming problems involve maximizing or minimizing an *objective function*, subject to a *set of constraints*.

1. In a linear programming problem there is always an _____ we try to maximize or _____ .

2. A linear programming problem is challenging because there is a set of _____ that prevents us from doing "whatever we want".

3. Linear programming belongs to the broader subject of _____ .

3.3   LINEAR PROGRAMMING—THE GEOMETRIC METHOD

The methods for solving linear programming problems are rather mechanical and can be learned without knowing exactly why and how. A computer spreadsheet program such as Microsoft Excel can also be used to solve small scale linear programming problems. In real world applications of linear programming, when thousands of thousands of variables are involved, a main frame computer will probably need to be employed. It is more important for a student of finite math to be able to recognize a linear programming problem when it is present, to be able to translate a word problem into a math system, and to know that most linear programming problems can be solved by computers once they are correctly set up.

We will solve a couple of simple problems just to get a feel for linear programming. We will place the emphasis on the setting up of a linear programming problem after that. Even though computers are becoming more and more powerful every day, no computer in the world is capable of reading a complex word problem and translating the problem into a well set up mathematical system.

Let's now do a review on how to graph a two-variable linear inequality. With two-variable linear programming problems, graphing inequalities is actually the bulk of the work.

Example 3.3.1. *Graph the linear inequality $x + y \leq 5$.*

Solution.

1. *First draw the line $x + y = 5$ in the coordinate system.*

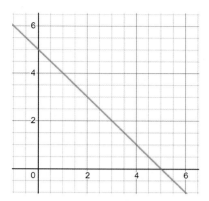

2. *Then pick a test point from either side of the line. Substitute the coordinates of this point for x and y in the inequality. If the inequality holds true after the substitution, then the side containing the test point is the solution; otherwise the other side is the solution. We leave the solution side blank and shade the other side. For this problem, we can pick the point $(1, 2)$ as our test point. Note that this point is below the line. Substitute 1 for x and 2 for y in the inequality and the inequality becomes $1 + 2 \leq 5$, which is a true statement. Therefore the side BELOW the line is the solution, and we shade the side ABOVE the line.*

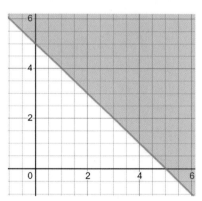

The graph is now complete.

For practice, please redo step 2 above with test points $(3, 4)$, $(3, 1)$ and $(0, 0)$. Note that a point such as $(4, 1)$ cannot be used as a test point because it is on the straight line itself. Also observe that $(0, 0)$ is always a good test point to use so long as it is not on the line◄

We are now ready to solve a linear programming problem.

Example 3.3.2. *Solve the linear programming problem from last section.*

$$\text{Maximize} \quad 3000x + 3500y$$

$$\text{Subject to} \quad \begin{cases} x + y \leq 5 \\ 6x + 9y \leq 36 \\ x \geq 0 \\ y \geq 0 \end{cases}$$

Solution. *Here are the steps we follow:*

1. *Graph the constraints. The final graph is a region in the x-y plane. This region is called the feasible region. Any point in the feasible region satisfies the constraints, and hence is a potential solution point.*
2. *The feasible region is bordered by line segments. Two line segments meet at a point that is called a vertex. We will find all vertices and their coordinates.*
3. *Finally we evaluate the objective function at each vertex to obtain a list of objective function values at all the vertices. If the problem calls for "Maximize objective function" then the maximum value in the list is the solution; if the problem calls for "Minimize objective function" then the minimum value in the list is the solution.*

Let's start.

1. *First we graph the inequality $x + y \leq 5$*

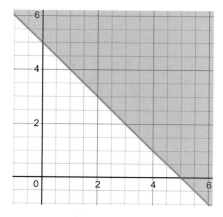

Then $6x + 9y \leq 36$

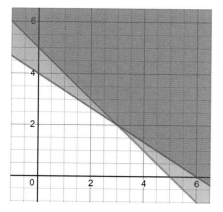

Then $x \geq 0$

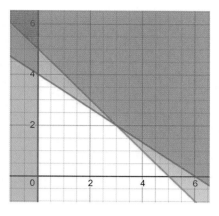

Then $y \geq 0$

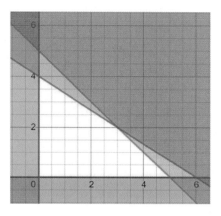

2. *The region that's never crossed out is the feasible region. We can see there are four vertices. Let's label them A, B, C and D as shown.*

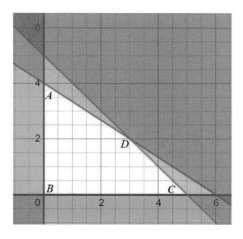

Vertex A has coordinates $(0, 4)$, vertex B has coordinates $(0, 0)$, vertex C has coordinates $(5, 0)$, and vertex D has coordinates $(3, 2)$. The coordinates of vertices A, B and C are easy to see because they are on the x- or y-axis. The coordinates of D can be found by solving the system

of the two equations whose graphs intersect at the point D:

$$\begin{cases} x + y = 5 \\ 6x + 9y = 36 \end{cases} \implies \text{Solution: } (3, 2)$$

3. *We now have the coordinates of all vertices and are ready for the final step.*

Vertex	Objective function $3000x + 3500y$
$A(0, 4)$	$3000 \times 0 + 3500 \times 4 = 14000$
$B(0, 0)$	$3000 \times 0 + 3500 \times 0 = 0$
$C(5, 0)$	$3000 \times 5 + 3500 \times 0 = 15000$
$D(3, 2)$	$3000 \times 3 + 3500 \times 2 = 16000$

The maximum value is 16000, *which occurs at the vertex* $D(3, 2)$, *i.e., Patrick should teach* 3 *classes at college X and* 2 *classes at college Y to achieve a maximum income of* $16,000 ◀

Please verify this: If we change the objective function in the example above to $3500x + 3000y$, then the maximum objective function value is $17,500 at vertex C.

Let's do another one. This time we will perform all the steps without the detailed explanations. By doing this you will see how the process is actually not as cumbersome as the last example has shown.

Example 3.3.3. *Solve the linear programming problem.*

$$\begin{aligned} \text{Minimize} \quad & 100 - 3x + 5y \\ \text{Subject to} \quad & \begin{cases} 2x + y \le 8 \\ x + 2y \le 10 \\ x - 3y \le -3 \\ x \ge 0 \\ y \ge 0 \end{cases} \end{aligned}$$

Solution. *We first graph the inequalities and find the vertices of the feasible region.*

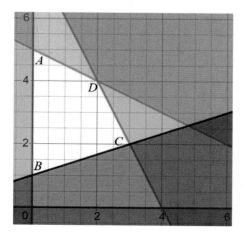

Vertex A has coordinates $(0,5)$, *vertex B has coordinates* $(0,1)$, *vertex C has coordinates* $(3,2)$, *and vertex D has coordinates* $(2,4)$. *The coordinates of C and D are obtained by solving the systems*

$$\begin{cases} 2x+ y = 8 \\ x-3y = -3 \end{cases} \quad and \quad \begin{cases} 2x+ y = 8 \\ x+2y = 10 \end{cases}$$

respectively. We then evaluate the objective function at the vertices.

Vertex	Objective function $100 - 3x + 5y$
$A(0,5)$	$100 - 3 \times 0 + 5 \times 5 = 125$
$B(0,1)$	$100 - 3 \times 0 + 5 \times 1 = 105$
$C(3,2)$	$100 - 3 \times 3 + 5 \times 2 = 101$
$D(2,4)$	$100 - 3 \times 2 + 5 \times 4 = 114$

The minimum objective function value is 101 *at vertex C* ◀

The process seems quite simple: (1) graph the feasible region, (2) find the vertices, (3) evaluate the objective function at all the vertices and DONE. But there are situations that will require a lot of work to fully investigate. For example, the feasible region may become unbounded and in such cases a solution does not necessarily exist; or sometimes there is more than one vertex that solves the problem; or maybe a solution does not make sense in the real world even though it is mathematically correct, such as when the solution says Patrick should teach 1.9 classes at college X and 3.1 classes at college Y. We will not venture into these difficult territories.

As mentioned earlier, no computer can set up a mathematical system from a real-life situation for us, yet. Computers are good at doing what we tell them to do, but we are the ones who need to know what we want or need to do, hence the importance of setting up mathematical systems.

The first step in setting up a system from a story problem is identifying all the variables involved. The second step is examining whether some of the variables are related to each other. The third step is expressing all the relations among the variables as described in the story in terms of mathematical functions, equations, and inequalities. Once all of that is done correctly, a beautiful system is set up, and from there computers can take over, most of the time anyway.

Example 3.3.4. *Set up a linear programming system for the following problem: A company is considering buying some new trucks. A model I truck has a payload capacity of* $1,100$ *pounds, costs* $\$30,000$, *incurs an operating cost of* $\$8,000$ *per year, and generates* 700 *pounds of* CO_2 *per year; a model II truck has a payload capacity of* $1,200$ *pounds, costs* $\$35,000$, *incurs an operating cost of* $\$7,000$ *per year, and generates* 600 *pounds of* CO_2 *per year; a model III truck has a payload capacity of* $1,300$ *pounds, costs* $\$40,000$, *incurs an operating cost of* $\$7,500$ *per year, and generates* 500 *pounds of* CO_2 *per year. The company has a purchasing budget of* $\$400,000$, *can afford* $\$100,000$ *in operating costs per year, and would like to generate no more than* $8,500$ *pounds of*

CO_2 per year. How many of each model should the company buy in order to maximize the total payload capacity?

Solution. *First we define the variables: Let x = number of model I trucks, y = number of model II trucks, and z = number of model III trucks.*

The numbers given in the sentences can be a little overwhelming, so let's organize them into tabular form.

	Model I	Model II	Model III	Conditions and Constraints
Payload	1100	1200	1300	To be maximized
Cost	30000	35000	40000	Cannot exceed 400000
Operating Cost	8000	7000	7500	Cannot exceed 100000
CO_2	700	600	500	Cannot exceed 8500

We can now easily see the total payload capacity of the fleet is $1100x + 1200y + 1300z$. The linear programming system is

$$\text{Maximize} \quad 1100x + 1200y + 1300z$$

$$\text{Subject to} \quad \begin{cases} 30000x + 35000y + 40000z \leq 400000 \\ 8000x + 7000y + 7500z \leq 100000 \\ 700x + 600y + 500z \leq 8500 \\ x, y, z \geq 0 \end{cases} \quad \blacktriangleleft$$

The system we just set up can no longer be solved by the geometric method because there are three variables involved, and the feasible region cannot be graphed in the x-y coordinate system. The Simplex Method is one of the methods that can be used to solve systems that involve three or more variables.

≡ EXERCISES 3.3 ≡

1. Match each inequality with a graph. Remember the region that is not shaded is the solution region.

   (A) $2x + 3y \leq 6$ (C) $3x - y \leq 3$ (E) $y \leq 2x + 1$ (G) $x \geq 0$

   (B) $2x + 3y \geq 6$ (D) $3x - y \geq 3$ (F) $y \geq (3/2)x$ (H) $y \geq 0$

White = solution

(i)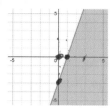

(iii)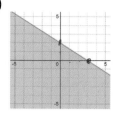

(v)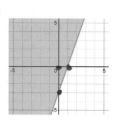

(vii)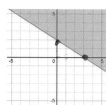

(ii)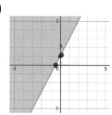

(iv)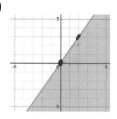

(vi)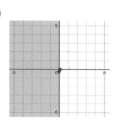

(viii)

2. Graph the inequalities.

   (a) $4x - 5y \leq 20$

   (b) $5x + 3.5y \geq 17.5$

   (c) $2x + 3y \geq 0$

   (d) $\begin{cases} 4x + 5y \leq 20 \\ 4x - 3y \leq 12 \\ x, y \geq 0 \end{cases}$

3. Solve the linear programming problem by following the steps: (i) Graph the set of inequalities. Use graph paper or a ruler to make the graphs as precise as possible. Rough sketches may result in getting an incorrect graph. (ii) Find the vertices of the feasible region. (iii) Evaluate the objective function at the vertices. (iv) State the optimal objective function value and the vertex at which the optimal value is obtained.

   (a) Maximize $7x + 3y$

   Subject to $\begin{cases} 2x + y \leq 8 \\ x + y \leq 6 \\ y \geq x - 1 \\ x, y \geq 0 \end{cases}$

   (b) Use the previous feasible region. Maximize $7x + 5y$.

   (c) Use the previous feasible region. Maximize $2x - 3y$.

   (d) Use the previous feasible region. Minimize $2x - 3y$.

   (e) Maximize $5x - 2y$

   Subject to $\begin{cases} 2x + y \geq 2 \\ 2x + 3y \leq 12 \\ y \leq -2x + 6 \\ x, y \geq 0 \end{cases}$

   (f) Use the previous feasible region. Maximize $3x + 4y$.

   (g) Use the previous feasible region. Minimize $3x - 4y + 25$.

4. Set up the following linear programming problem: A man designs a monthly exercise program consisting of shadow boxing, rope jumping, and icy water bathing. He would like to exercise at most 60 hours, devote at most 10 hours to icy water bathing, and jump rope for no more than the total number of hours shadow boxing and icy water bathing combined. The calories burned per hour by shadow boxing, rope jumping, and icy water bathing are 300, 500, and 50, respectively. How many hours should be allotted to each activity to maximize the number of calories burned?

   (a) Use three letters to represent the three unknown quantities to be found.

   (b) Find the objective function, and clearly indicate whether it should be maximized or minimized.

   (c) List all the constraints.

5. Set up the following linear programming problem: A Meals-On-Wheels business sells three ultra-popular lunch dishes. Dish A requires 2 minutes of preparation, 3 minutes of cooking, 1 minute of packaging, and generates a profit of $5 per item; dish B requires 4 minutes of preparation, 0 minutes of cooking, 2 minutes of packaging, and generates a profit of $7 per item; dish C requires 5 minutes of preparation, 5 minutes of cooking, 3 minutes of packaging, and generates a profit of $12 per item. The business has a maximum of 110 minutes of preparation time, 80 minutes of cooking time, and 60 minutes of packaging time available per day. How many of each dish should the business prepare per day to maximize the profit?

   (a) Use three letters to represent the three unknown quantities to be found.

   (b) Find the objective function, and clearly indicate whether it should be maximized or minimized.

   (c) List all the constraints.

Meals on Wheels

6. Set up the following linear programming problem: A diet requires meeting a minimum of 500 units of nutrient I, 400 units of nutrient II, 300 units of nutrient III, and 200 units of nutrient IV. There are five foods available, their nutrient contents and costs per pound are given below. Find the amount of each food needed in order to meet the nutrient requirements with the minimum cost.

	Nutrient I	Nutrient II	Nutrient III	Nutrient IV	Cost
Food V	a_{11}	a_{21}	a_{31}	a_{41}	c_V
Food W	a_{12}	a_{22}	a_{32}	a_{42}	c_W
Food X	a_{13}	a_{23}	a_{33}	a_{43}	c_X
Food Y	a_{14}	a_{24}	a_{34}	a_{44}	c_Y
Food Z	a_{15}	a_{25}	a_{35}	a_{45}	c_Z

(a) Use five letters to represent the five unknown quantities to be found.

(b) Find the objective function, and clearly indicate whether it should be maximized or minimized.

(c) List all the constraints.

7. Set up the following linear programming problem: Moonshine Nucky needs to make 800 gallons of his special Triple Shine Moonshine that's at least 50% but no more than 60% alcohol by volume. He mixes three moonshines from three different suppliers. Supplier A's stuff is 47% alcohol and costs $5 per gallon; supplier B's is 55% alcohol and costs $4 per gallon; and supplier C's is 80% alcohol and costs $3 per gallon. To maintain a good business relationship with all his suppliers, Nucky must purchase at least 100 gallons but no more than 400 gallons from each supplier. How many gallons of the good stuff must Nucky purchase from each supplier in order to create his product at the minimum cost?

8. Set up the following linear programming problem: After years of working hard making money, Moonshine Nucky is trying to give back to the community. He plans to establish a Food & Roof Foundation to help the less fortunate. Building the Foundation's main building requires 500 tons of timber and 900 tons of stone. Two companies in the area supply timber and stone. Company A charges $5/ton for timber and $4/ton for stone and a minimum of 200 tons of material must be ordered. Company B charges $6/ton for timber and $2/ton for stone and a minimum of 250 tons of material must be ordered. How many tons of timber and how many tons of stone must Nucky order from each company to minimize the total material cost?

9. Can today's computers read a story problem like the one above and proceed to set up a linear programming problem for us?

3.4 MATRIX OPERATIONS

Matrices have a wide range of applications. We have seen one in the Gauss-Jordan elimination method. The simplest application is probably the storage of data. If you take a movie poster you no longer want, cut it vertically and horizontally into, say, 40,000 small squares, each square will consist of roughly one color. You may now assign a number to a color, and when you put all the numbers together according to where each square belongs on the poster, what you get is a MATRIX. That is actually how digital cameras work (more or less). Read a digital camera's description and you will see pixel numbers. The numbers are telling you into how many small pieces the camera cuts each picture (not quite that simple but good enough for a layman like me). The higher the pixel count, the better the picture quality. The camera sees a picture, cuts it into pieces, assigns a number to each piece based on the color of that piece, and when it is time to display the picture, the camera translates the numbers back to colors, and you have a picture—you can test this by opening a picture on your computer screen and continuing to zoom in. Below is an oversimplified example, in which an 8 by 8 matrix is translated to an 8 by 8 table by converting a 0 to no color, a 1 to green, a 2 to yellow, and a 3 to red in the corresponding cell.

$$\begin{bmatrix} 0 & 0 & 0 & 0 & 0 & 0 & 0 & 0 \\ 0 & 0 & 0 & 0 & 0 & 0 & 0 & 0 \\ 0 & 1 & 1 & 0 & 0 & 2 & 2 & 0 \\ 0 & 1 & 1 & 0 & 0 & 2 & 2 & 0 \\ 0 & 0 & 0 & 0 & 0 & 0 & 0 & 0 \\ 0 & 0 & 0 & 0 & 0 & 0 & 0 & 0 \\ 0 & 0 & 3 & 3 & 3 & 3 & 0 & 0 \\ 0 & 0 & 0 & 0 & 0 & 0 & 0 & 0 \end{bmatrix}$$

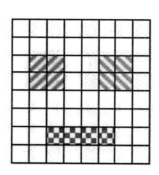

green

yellow

red

If you step back far enough, the colors in the squares kind of form a face with one green eye, one yellow eye, and red lips.

We will cover some basic matrix operations here, just to get a taste, and appreciate how matrices are used in some applications.

First some jargon.

- The *size of a matrix* is described by the number of rows and the number of columns in the matrix. The following are three matrices with various sizes.

$$\begin{bmatrix} a & b & c & d \\ 1 & 2 & 3 & 4 \\ x & y & z & u \end{bmatrix} \text{ is a } 3 \times 4, \quad \begin{bmatrix} 10 \\ 9 \\ 8 \end{bmatrix} \text{ is a } 3 \times 1, \text{ and } \begin{bmatrix} a & b & c & d \end{bmatrix} \text{ is a } 1 \times 4 \text{ matrix.}$$

- A number in a matrix is called an *element* or an *entry*.

- An element in a matrix can be referred to by its location in the matrix. For example, the number 5 is the 2nd row-3rd column element in the following matrix.

$$\begin{bmatrix} 1 & 0 & 0 & 0 \\ 0 & 2 & 5 & 0 \\ 0 & 0 & 3 & 0 \end{bmatrix}$$

- An $m \times n$ matrix can be written in the general form

$$
\begin{bmatrix}
a_{11} & a_{12} & a_{13} & \cdots & a_{1n} \\
a_{21} & a_{22} & a_{23} & \cdots & a_{2n} \\
a_{31} & a_{32} & a_{33} & \cdots & a_{3n} \\
\vdots & \vdots & \vdots & \ddots & \vdots \\
a_{m1} & a_{m2} & a_{m3} & \cdots & a_{mn}
\end{bmatrix}
$$

The 2nd row-3rd column element is a_{23}. Though it may look a little cumbersome at first glance, this notation is very convenient in mathematics expressions as it saves us from using too many words to describe or explain things. Look at the matrix again until you figure out how the subscripts are arranged. Notations are a big part of math—they are the language of math. We cannot do science without math, and we cannot do math without math notations.

- Two matrices are equal if they have the same size and same corresponding elements. For example, if we know that

$$
\begin{bmatrix}
x \\
y \\
z
\end{bmatrix}
=
\begin{bmatrix}
3 \\
4 \\
-5
\end{bmatrix}
$$

then $x = 3$, $y = 4$, $z = -5$.

Suppose k is a real number, A is an $m \times n$ matrix, and B is a $p \times q$ matrix:

$$
A =
\begin{bmatrix}
a_{11} & a_{12} & a_{13} & \cdots & a_{1n} \\
a_{21} & a_{22} & a_{23} & \cdots & a_{2n} \\
a_{31} & a_{32} & a_{33} & \cdots & a_{3n} \\
\vdots & \vdots & \vdots & \ddots & \vdots \\
a_{m1} & a_{m2} & a_{m3} & \cdots & a_{mn}
\end{bmatrix}
\quad \text{and} \quad
B =
\begin{bmatrix}
b_{11} & b_{12} & b_{13} & \cdots & b_{1q} \\
b_{21} & b_{22} & b_{23} & \cdots & b_{2q} \\
b_{31} & b_{32} & b_{33} & \cdots & b_{3q} \\
\vdots & \vdots & \vdots & \ddots & \vdots \\
b_{p1} & b_{p2} & b_{p3} & \cdots & b_{pq}
\end{bmatrix}
$$

We now define the basic matrix operations.

1. *Scalar multiplication* is the product of a real number and a matrix.

$$
kA =
\begin{bmatrix}
ka_{11} & ka_{12} & ka_{13} & \cdots & ka_{1n} \\
ka_{21} & ka_{22} & ka_{23} & \cdots & ka_{2n} \\
ka_{31} & ka_{32} & ka_{33} & \cdots & ka_{3n} \\
\vdots & \vdots & \vdots & \ddots & \vdots \\
ka_{m1} & ka_{m2} & ka_{m3} & \cdots & ka_{mn}
\end{bmatrix}
$$

2. *Matrix addition* is the sum of two matrices of the same size. The addition of two matrices

of different sizes is undefined. If A and B are both $m \times n$, then

$$A + B = \begin{bmatrix} a_{11} + b_{11} & a_{12} + b_{12} & a_{13} + b_{13} & \cdots & a_{1n} + b_{1n} \\ a_{21} + b_{21} & a_{22} + b_{22} & a_{23} + b_{23} & \cdots & a_{2n} + b_{2n} \\ a_{31} + b_{31} & a_{32} + b_{32} & a_{33} + b_{33} & \cdots & a_{3n} + b_{3n} \\ \vdots & \vdots & \vdots & \ddots & \vdots \\ a_{m1} + b_{m1} & a_{m2} + b_{m2} & a_{m3} + b_{m3} & \cdots & a_{mn} + b_{mn} \end{bmatrix}$$

3. *Matrix multiplication* is the product of two matrices, and is a little more "unnatural" than the first two operations. After seeing how matrix addition is defined, most people would expect matrix multiplication to be similarly defined by simply changing the "add" operation to the "multiply" operation. This, however, turns out to be not as useful (a loaded word here, do some research if you are intrigued). We will now proceed to the formal definition.

If A is an $m \times n$ matrix and B is an $n \times q$ matrix, then AB is an $m \times q$ matrix whose i^{th} row-j^{th} column element equals $a_{i1}b_{1j} + a_{i2}b_{2j} + \cdots + a_{in}b_{nj}$, where $i = 1, 2, \ldots, m$ and $j = 1, 2, \ldots, q$. In other words, we "match" each and every row of A with each and every column of B to create the product AB. For this to happen, the number of columns in A must equal the number of rows in B.

4. If multiple operations are involved, multiplications precede additions, and operations inside parentheses precede the ones outside.

Let's look at some examples.

Example 3.4.1. *A scalar multiplication.*

$$3 \begin{bmatrix} -1 & 0 \\ 2 & 4 \\ 0 & 5 \end{bmatrix} = \begin{bmatrix} -3 & 0 \\ 6 & 12 \\ 0 & 15 \end{bmatrix} \blacktriangleleft$$

Example 3.4.2. *A matrix addition.*

$$\begin{bmatrix} -1 & 0 \\ 2 & 4 \\ 0 & 5 \end{bmatrix} + \begin{bmatrix} -3 & 0 \\ 6 & 12 \\ 0 & 15 \end{bmatrix} = \begin{bmatrix} -4 & 0 \\ 8 & 16 \\ 0 & 20 \end{bmatrix} \blacktriangleleft$$

Example 3.4.3. *An undefined matrix addition.*

$$\begin{bmatrix} -1 & 0 \\ 2 & 4 \\ 0 & 5 \end{bmatrix} + \begin{bmatrix} 3 \\ 7 \\ 8 \end{bmatrix}$$

is undefined because the two matrices are of different sizes $\blacktriangleleft$

Example 3.4.4. *A matrix multiplication. Read this one slowly to make sure you see everything. Given two matrices*

$$A = \begin{bmatrix} -1 & 3 \\ 2 & 4 \\ 6 & 5 \end{bmatrix} \text{ and } B = \begin{bmatrix} 1 & 2 & 3 & 4 \\ 2 & 3 & 4 & 5 \end{bmatrix}$$

(a) *Determine whether AB is defined.*
(b) *Find AB if it is defined.*

Solution.

(a) *A is 3×2, B is 2×4, i.e., A has 2 columns and B has 2 rows, $2 = 2$, the product AB is defined. We also know that AB is 3×4.*

(b) *We "match" the first row of A with the first column of B by taking the product of their first numbers, then the product of their second numbers, and adding the two products together: $(-1)(1) + (3)(2) = 5$. This will be the first row-first column element of AB:*

$$\begin{bmatrix} \boxed{-1} & \boxed{3} \\ 2 & 4 \\ 6 & 5 \end{bmatrix} \begin{bmatrix} \boxed{1} & 2 & 3 & 4 \\ \boxed{2} & 3 & 4 & 5 \end{bmatrix} = \begin{bmatrix} \boxed{5} & & & \\ & & & \\ & & & \end{bmatrix}$$

We then "match" the first row of A with the second column of B to obtain the first row-second column element of AB:

$$\begin{bmatrix} \boxed{-1} & \boxed{3} \\ 2 & 4 \\ 6 & 5 \end{bmatrix} \begin{bmatrix} 1 & \boxed{2} & 3 & 4 \\ 2 & \boxed{3} & 4 & 5 \end{bmatrix} = \begin{bmatrix} 5 & \boxed{7} & & \\ & & & \\ & & & \end{bmatrix}$$

Move on to the first row of A and the third column of B:

$$\begin{bmatrix} \boxed{-1} & \boxed{3} \\ 2 & 4 \\ 6 & 5 \end{bmatrix} \begin{bmatrix} 1 & 2 & \boxed{3} & 4 \\ 2 & 3 & \boxed{4} & 5 \end{bmatrix} = \begin{bmatrix} 5 & 7 & \boxed{9} & \\ & & & \\ & & & \end{bmatrix}$$

Now the first row of A and the fourth column of B:

$$\begin{bmatrix} \boxed{-1} & \boxed{3} \\ 2 & 4 \\ 6 & 5 \end{bmatrix} \begin{bmatrix} 1 & 2 & 3 & \boxed{4} \\ 2 & 3 & 4 & \boxed{5} \end{bmatrix} = \begin{bmatrix} 5 & 7 & 9 & \boxed{11} \\ & & & \\ & & & \end{bmatrix}$$

All columns of B have now been "matched" with the first row of A to generate the first row of AB. The second row of AB is generated by "matching" the second row of A with all columns of B:

$$\begin{bmatrix} -1 & 3 \\ \boxed{2} & \boxed{4} \\ 6 & 5 \end{bmatrix} \begin{bmatrix} \boxed{1} & 2 & 3 & 4 \\ \boxed{2} & 3 & 4 & 5 \end{bmatrix} = \begin{bmatrix} 5 & 7 & 9 & 11 \\ \boxed{10} & 16 & 22 & 28 \end{bmatrix}$$

Finally the third row of AB is generated by "matching" the third row of A with all columns of B:

$$\begin{bmatrix} -1 & 3 \\ 2 & 4 \\ \boxed{6} & \boxed{5} \end{bmatrix} \begin{bmatrix} \boxed{1} & 2 & 3 & 4 \\ \boxed{2} & 3 & 4 & 5 \end{bmatrix} = \begin{bmatrix} 5 & 7 & 9 & 11 \\ 10 & 16 & 22 & 28 \\ \boxed{16} & 27 & 38 & 49 \end{bmatrix} \blacktriangleleft$$

Example 3.4.5. *Given two matrices*

$$A = \begin{bmatrix} -1 & 3 \\ 2 & 4 \\ 6 & 5 \end{bmatrix} \ and \ B = \begin{bmatrix} 1 & 2 & 3 & 4 \\ 2 & 3 & 4 & 5 \end{bmatrix}$$

Determine whether BA is defined.

Solution. *You might have noticed that these two matrices are the two from the previous example, but here we are looking at BA instead of AB. Unlike the multiplication of two numbers, which is commutative: $3 \cdot 4 = 4 \cdot 3$, matrix multiplication is not commutative. When AB is defined, BA may be (a) undefined, or (b) defined but not equal to AB, or (c) defined and equal to AB.*

For this particular problem, BA is undefined because B has 4 columns (B is 2×4), and A has 3 rows (A is 3×2), $4 \neq 3$. If we try anyway, we will see the problem:

$$\begin{bmatrix} \boxed{1} & \boxed{2} & \boxed{3} & \boxed{4} \\ 2 & 3 & 4 & 5 \end{bmatrix} \begin{bmatrix} \boxed{-1} & 3 \\ \boxed{2} & 4 \\ \boxed{6} & 5 \end{bmatrix}$$

We see that every row of B has 4 elements, while every column of A has 3 elements, we are unable to "match" their elements one-to-one◀

Example 3.4.6. *Suppose A is a 2×4, B is a 4×3, C is a 4×2 and D is a 2×3 matrix. Determine whether each of the following operations is defined, and if it is, the size of the final matrix.*

(a) *ABC.*

(b) *CAB.*

(c) *$AB + D$.*

Solution.

(a) *AB is a 2×3 and C is a 4×2. $3 \neq 4$, ABC is undefined.*

(b) *CA is a 4×4 and B is a 4×3. CAB is a 4×3.*

(c) *AB is a* 2 × 3 *and D is a* 2 × 3. *Their sum AB + D is also a* 2 × 3 ◀

Example 3.4.7. *This example introduces two new definitions: the identity matrix and the inverse matrix. Given three matrices:*

$$A = \begin{bmatrix} 1 & -2 & 3 \\ 2 & -3 & 4 \\ -2 & -1 & 1 \end{bmatrix} \qquad B = \begin{bmatrix} -1/3 & 1/3 & -1/3 \\ 10/3 & -7/3 & -2/3 \\ 8/3 & -5/3 & -1/3 \end{bmatrix} \qquad and \qquad I = \begin{bmatrix} 1 & 0 & 0 \\ 0 & 1 & 0 \\ 0 & 0 & 1 \end{bmatrix}$$

It is easy to verify the following results:

(a) $IA = A$.

(b) $BI = B$.

(c) $AB = I$.

(d) $BA = I$ ◀

As we can see from this example, the structure of I is easy to identify. It has 1's along the upper left to lower right diagonal and 0's elsewhere. Multiplying any matrix by I, either on the left or on the right, does not change the matrix, as long as the multiplication is defined. The matrix I is called the *identity matrix*. If necessary, a subscript can be used to identify its size: I_2 is the 2 × 2 identity matrix, I_3 is the 3 × 3 identity matrix, etc.

Matrices A and B are uniquely related in that their products AB and BA both equal the identity matrix. They are said to be *inverse matrices* of each other. B is called the *inverse matrix* of A and can be written as $B = A^{-1}$.

Not every matrix has an inverse matrix. If a matrix has an inverse, the inverse can be found by the Gauss-Jordan elimination method. We will demonstrate the method through an example at the end of this section. A graphing calculator can be used to find the inverse matrix of a small matrix. Large inverse matrices (500 by 500 for example) are found by using computer programs.

The inverse matrix can be used to solve an $n \times n$ system of linear equations.

Example 3.4.8. *Use the inverse matrix to solve the following system.*

$$\begin{cases} x & - & 2y & + & 3z & = & -9 \\ 2x & - & 3y & + & 4z & = & 6 \\ -2x & - & y & + & z & = & 12 \end{cases}$$

Solution. *We know the system can be solved by using the Gauss-Jordan method. We will take a new approach here and explain why this approach is more desirable in some situations. First, we* *write the system as a matrix equation:*

$$\begin{bmatrix} 1 & -2 & 3 \\ 2 & -3 & 4 \\ -2 & -1 & 1 \end{bmatrix} \begin{bmatrix} x \\ y \\ z \end{bmatrix} = \begin{bmatrix} -9 \\ 6 \\ 12 \end{bmatrix}$$

Notice that the left-hand side is the product of a 3 × 3 matrix and a 3 × 1 matrix, making it a 3 × 1 matrix. You should perform the multiplication to gain a clear view of how this matrix equation is equivalent to the original system. Make sure you see where the numbers in the matrix equation come from.

Once you see that this matrix equation represents the original system, the rest is easy. The inverse of the matrix

$$\begin{bmatrix} 1 & -2 & 3 \\ 2 & -3 & 4 \\ -2 & -1 & 1 \end{bmatrix}$$

is (we will see how this inverse matrix is found later)

$$\begin{bmatrix} -1/3 & 1/3 & -1/3 \\ 10/3 & -7/3 & -2/3 \\ 8/3 & -5/3 & -1/3 \end{bmatrix}$$

We multiply both sides of the equation by the inverse matrix:

$$\begin{bmatrix} -1/3 & 1/3 & -1/3 \\ 10/3 & -7/3 & -2/3 \\ 8/3 & -5/3 & -1/3 \end{bmatrix} \begin{bmatrix} 1 & -2 & 3 \\ 2 & -3 & 4 \\ -2 & -1 & 1 \end{bmatrix} \begin{bmatrix} x \\ y \\ z \end{bmatrix} = \begin{bmatrix} -1/3 & 1/3 & -1/3 \\ 10/3 & -7/3 & -2/3 \\ 8/3 & -5/3 & -1/3 \end{bmatrix} \begin{bmatrix} -9 \\ 6 \\ 12 \end{bmatrix}$$

After one multiplication on each side, we get:

$$\begin{bmatrix} 1 & 0 & 0 \\ 0 & 1 & 0 \\ 0 & 0 & 1 \end{bmatrix} \begin{bmatrix} x \\ y \\ z \end{bmatrix} = \begin{bmatrix} 1 \\ -52 \\ -38 \end{bmatrix}$$

The identity matrix on the left-hand side doesn't change the column matrix, so:

$$\begin{bmatrix} x \\ y \\ z \end{bmatrix} = \begin{bmatrix} 1 \\ -52 \\ -38 \end{bmatrix}$$

The solution is (1,−52,−38) If we go back and go through the entire process again, we see that the solution is obtained by finding the product of the inverse matrix of the system matrix and the right-hand side column matrix◄

Let's do one more example.

Example 3.4.9. *Solve the following system.*

$$\begin{cases} x & - & 3y & + & 4z & = & -6 \\ 3x & + & y & + & 9z & = & -2 \\ 4x & - & y & - & z & = & 21 \end{cases}$$

Solution. *First, we write the system as a matrix equation:*

$$\begin{bmatrix} 1 & -3 & 4 \\ 3 & 1 & 9 \\ 4 & -1 & -1 \end{bmatrix} \begin{bmatrix} x \\ y \\ z \end{bmatrix} = \begin{bmatrix} -6 \\ -2 \\ 21 \end{bmatrix}$$

Next, we find the inverse of the system matrix:

$$\begin{bmatrix} -8/137 & 7/137 & 31/137 \\ -39/137 & 17/137 & -3/137 \\ 7/137 & 11/137 & -10/137 \end{bmatrix}$$

Now find the product of this inverse matrix and the column matrix on the right-hand side to obtain the solution:

$$\begin{bmatrix} x \\ y \\ z \end{bmatrix} = \begin{bmatrix} -8/137 & 7/137 & 31/137 \\ -39/137 & 17/137 & -3/137 \\ 7/137 & 11/137 & -10/137 \end{bmatrix} \begin{bmatrix} -6 \\ -2 \\ 21 \end{bmatrix} = \begin{bmatrix} 5 \\ 1 \\ -2 \end{bmatrix}$$

The solution is $(5, 1, -2)$ ◄

Example 3.4.10. *A company makes two styles of jeans. A Classic sells for $60 each, a Slim Fit sells for $80 each. The jeans are sold at two outlets. Last month outlet A sold 36 pairs of jeans for $2460, and outlet B sold 52 pairs for $3720. How many pairs of each style were sold at each outlet last month?*

Solution. *Let's consider outlet A first. Suppose x pairs of Classic were sold and y pairs of Slim Fit were sold. The total number sold is* $x + y$, *and the revenue is* $60x + 80y$. *We have the following system for outlet A:*

$$\begin{cases} x & + & y & = & 36 \\ 60x & + & 80y & = & 2460 \end{cases}$$

Similarly, the system for outlet B is:

$$\begin{cases} x & + & y & = & 52 \\ 60x & + & 80y & = & 3720 \end{cases}$$

We could solve each system using the Gauss-Jordan method, but that would be almost like going through the same process twice because the left-hand sides of the two systems are identical. If we use the inverse matrix, then we only need to find the inverse matrix once and use it to solve both systems. The inverse matrix method is even better if the company sells its jeans at several places.

The matrix associated with the system and its inverse are given below:

$$\text{System matrix} = \begin{bmatrix} 1 & 1 \\ 60 & 80 \end{bmatrix} \qquad \text{Inverse matrix} = \begin{bmatrix} 4 & -1/20 \\ -3 & 1/20 \end{bmatrix}$$

The solution to outlet A is

$$\begin{bmatrix} 4 & -1/20 \\ -3 & 1/20 \end{bmatrix} \begin{bmatrix} 36 \\ 2460 \end{bmatrix} = \begin{bmatrix} 21 \\ 15 \end{bmatrix}$$

The solution to outlet B is

$$\begin{bmatrix} 4 & -1/20 \\ -3 & 1/20 \end{bmatrix} \begin{bmatrix} 52 \\ 3720 \end{bmatrix} = \begin{bmatrix} 22 \\ 30 \end{bmatrix}$$

I.e., 21 pairs of Classic and 15 pairs of Slim Fit were sold at outlet A; and 22 pairs of Classic and 30 pairs of Slim Fit were sold at outlet B ◀

As promised, we will show, before we close this section, how to find the inverse matrix of a matrix that has an inverse. The example will satisfy some people's curiosity, and arouse some other people's greater curiosity so that they will go on to do more investigation.

Example 3.4.11. *Find the inverse matrix of the matrix*

$$A = \begin{bmatrix} 1 & -2 & 3 \\ 2 & -3 & 4 \\ -2 & -1 & 1 \end{bmatrix}$$

Solution. *The process involves putting a 3×3 identity matrix right next to the given matrix to form a 3×6 matrix, then performing row operations to convert the first three columns into an identity matrix. The last three columns will then form the inverse matrix.*

$$\left[\begin{array}{ccc|ccc} 1 & -2 & 3 & 1 & 0 & 0 \\ 2 & -3 & 4 & 0 & 1 & 0 \\ -2 & -1 & 1 & 0 & 0 & 1 \end{array} \right]$$

$$\begin{array}{c} -2R1 + R2 \\ 2R1 + R3 \\ \xrightarrow{\hspace{2cm}} \end{array} \left[\begin{array}{ccc|ccc} 1 & -2 & 3 & 1 & 0 & 0 \\ 0 & 1 & -2 & -2 & 1 & 0 \\ 0 & -5 & 7 & 2 & 0 & 1 \end{array} \right]$$

$$\begin{array}{c} 2R2 + R1 \\ 5R2 + R3 \\ \xrightarrow{\hspace{2cm}} \end{array} \left[\begin{array}{ccc|ccc} 1 & 0 & -1 & -3 & 2 & 0 \\ 0 & 1 & -2 & -2 & 1 & 0 \\ 0 & 0 & -3 & -8 & 5 & 1 \end{array} \right]$$

$$-\frac{1}{3}R3 \xrightarrow{} \left[\begin{array}{ccc|ccc} 1 & 0 & -1 & -3 & 2 & 0 \\ 0 & 1 & -2 & -2 & 1 & 0 \\ 0 & 0 & 1 & 8/3 & -5/3 & -1/3 \end{array}\right]$$

$$\begin{array}{c} R3+R1 \\ 2R3+R2 \end{array} \xrightarrow{} \left[\begin{array}{ccc|ccc} 1 & 0 & 0 & -1/3 & 1/3 & -1/3 \\ 0 & 1 & 0 & 10/3 & -7/3 & -2/3 \\ 0 & 0 & 1 & 8/3 & -5/3 & -1/3 \end{array}\right]$$

The first three columns now form the identity matrix, the process is complete, and the last three columns form the inverse matrix:

$$A^{-1} = \left[\begin{array}{ccc} -1/3 & 1/3 & -1/3 \\ 10/3 & -7/3 & -2/3 \\ 8/3 & -5/3 & -1/3 \end{array}\right] \blacktriangleleft$$

≡ **EXERCISES 3.4** ≡

1. State the size of each matrix.

(a) $\left[\begin{array}{ccc} 1 & 6 & 9 \\ 1 & 3 & 8 \end{array}\right]$ 2×3

(b) $\left[\begin{array}{cc} 1 & 1 \\ 6 & 3 \\ 9 & 8 \end{array}\right]$ 3×2

(c) $\left[\begin{array}{cccc} x & y & z & 5 \end{array}\right]$ 1×4

(d) $\left[\begin{array}{c} a \\ b \\ c \end{array}\right]$ 3×1

(e) $\left[\begin{array}{c} 7 \end{array}\right]$ 1×1

2. Given the matrix $\left[\begin{array}{ccc} 3 & -5 & 0 \\ 2 & 1 & -9 \end{array}\right]$

(a) The 1st row-2nd column element is ___−5___.

(b) The 2nd row-1st column element is ___2___.

(c) −9 is the ___2___ row-___3___ column element.

3. Given that the following two matrices are equal, find the values of x, y, z, A, B, C.

$x-1$ $A=2$
$y=4$ $B=3$
$z=5$ $C=6$

$$\left[\begin{array}{cc} x & 2 \\ 3 & y \\ z & 6 \end{array}\right] = \left[\begin{array}{cc} 1 & A \\ B & 4 \\ 5 & C \end{array}\right]$$

4. Perform the operations. Give a reason if an operation is undefined.

(a) $\begin{bmatrix} 2 & -4 \\ -3 & 5 \end{bmatrix} + \begin{bmatrix} 0 & -1 \\ 6 & -7 \end{bmatrix}$ [

(b) $2\begin{bmatrix} 0 & -1 \\ 6 & -7 \end{bmatrix}$

(c) $2 + \begin{bmatrix} 0 & -1 \\ 6 & -7 \end{bmatrix}$ undefined, cant add a # to matrix

(d) $\begin{bmatrix} 2 \end{bmatrix} + \begin{bmatrix} 0 & -1 \\ 6 & -7 \end{bmatrix}$ undefined— cant add matrices on diff sizes

(e) $3\begin{bmatrix} 2 \\ -4 \\ 6 \end{bmatrix} + \begin{bmatrix} 1 \\ 0 \\ -10 \end{bmatrix}$

(f) $4\begin{bmatrix} 2 & -4 \\ -3 & 5 \end{bmatrix} - 3\begin{bmatrix} 0 & -1 \\ 6 & -7 \end{bmatrix}$

(g) $\begin{bmatrix} 2 & 1 & 0 \\ -1 & 0 & 5 \\ 0 & 3 & 4 \end{bmatrix}\begin{bmatrix} 10 & -2 \\ 7 & 6 \\ -8 & 0 \end{bmatrix}$ 3×3 3×2

(h) $\begin{bmatrix} 1 & 0 & 1 & 0 \\ 0 & -1 & 0 & -1 \\ 0 & 0 & 0 & 0 \\ 1 & 3 & 5 & 7 \\ 2 & 4 & 6 & 8 \end{bmatrix}\begin{bmatrix} 1 & 2 & 3 \\ 4 & 5 & 6 \\ 7 & 8 & 9 \\ 0 & 0 & 0 \end{bmatrix}$ 5×3
5×4 4×3

undefined because diff sizes

(i) $\begin{bmatrix} 1 & 2 & 3 \\ 4 & 5 & 6 \\ 7 & 8 & 9 \\ 0 & 0 & 0 \end{bmatrix}\begin{bmatrix} 1 & 0 & 1 & 0 \\ 0 & -1 & 0 & -1 \\ 0 & 0 & 0 & 0 \\ 1 & 3 & 5 & 7 \\ 2 & 4 & 6 & 8 \end{bmatrix}$
4×3 5×4

(j) $\begin{bmatrix} -2 & 3 & 5 \\ 11 & 9 & 13 \\ 7 & -8 & 16 \end{bmatrix}\begin{bmatrix} x \\ y \\ z \end{bmatrix}$
3×3 3×1

(k) $\begin{bmatrix} 1 & 0 & 0 \\ 0 & 1 & 0 \\ 0 & 0 & 1 \end{bmatrix}\begin{bmatrix} x \\ y \\ z \end{bmatrix} = \begin{bmatrix} x \\ y \\ z \end{bmatrix}$

(l) $\begin{bmatrix} 0 & 0 & 1 \\ 0 & 1 & 0 \\ 1 & 0 & 0 \end{bmatrix}\begin{bmatrix} x \\ y \\ z \end{bmatrix} = \begin{bmatrix} z \\ y \\ x \end{bmatrix}$

(m) $\begin{bmatrix} 1 & 0 & 0 \\ 0 & 0 & 1 \\ 0 & 1 & 0 \end{bmatrix} \begin{bmatrix} x \\ y \\ z \end{bmatrix}$

(n) See how to shuffle three letters now?

(o) Skip this one because an o looks like a 0.

(p) $\begin{bmatrix} 2 & 4 & 6 \end{bmatrix} \begin{bmatrix} 1 \\ 3 \\ 5 \end{bmatrix}$

(q) $\begin{bmatrix} 2 & 4 & 6 \end{bmatrix} \begin{bmatrix} x \\ y \\ z \end{bmatrix}$ $[2x + 4y + 3z]$

(r) $\begin{bmatrix} 2 \\ 4 \\ 6 \end{bmatrix} \begin{bmatrix} 1 & 3 & 5 \end{bmatrix}$

(s) $\begin{bmatrix} 2 \\ 4 \\ 6 \end{bmatrix} \begin{bmatrix} x & y & z \end{bmatrix}$

(t) $\begin{bmatrix} 2 \\ 4 \\ 6 \end{bmatrix} \begin{bmatrix} x \\ y \\ z \end{bmatrix}$ undefined

(u) $\begin{bmatrix} 2 & 4 & 6 \end{bmatrix} \begin{bmatrix} x & y & z \end{bmatrix}$ undefined

(v) $\begin{bmatrix} 2 & 1 & 0 \\ -1 & 0 & 5 \\ 0 & 3 & 4 \end{bmatrix} \begin{bmatrix} 10 & -2 \\ 7 & 6 \\ -8 & 0 \end{bmatrix} + \begin{bmatrix} 3 & 8 \\ 30 & 28 \\ 51 & 12 \end{bmatrix}$
3x3   3x2

(w) $\begin{bmatrix} -2 & 3 & 5 \\ 11 & 9 & 13 \\ 7 & -8 & 16 \end{bmatrix} \begin{bmatrix} x \\ y \\ z \end{bmatrix} - \begin{bmatrix} 1 \\ 2 \\ 3 \end{bmatrix}$

5. Given a 2 × 3 matrix $\begin{bmatrix} 6 & 8 & 9 \\ 8 & 7 & 4 \end{bmatrix}$

(a) Find $\begin{bmatrix} 1 & 1 \end{bmatrix} \begin{bmatrix} 6 & 8 & 9 \\ 8 & 7 & 4 \end{bmatrix}$ and explain why this can be called the column sums.

(b) Find $\dfrac{1}{2} \begin{bmatrix} 1 & 1 \end{bmatrix} \begin{bmatrix} 6 & 8 & 9 \\ 8 & 7 & 4 \end{bmatrix}$ and explain why this can be called the column averages.

(c) Find $\begin{bmatrix} 6 & 8 & 9 \\ 8 & 7 & 4 \end{bmatrix} \begin{bmatrix} 1 \\ 1 \\ 1 \end{bmatrix}$ and expalin why this can be called the row sums.

(d) Find $\dfrac{1}{3}\begin{bmatrix} 6 & 8 & 9 \\ 8 & 7 & 4 \end{bmatrix}\begin{bmatrix} 1 \\ 1 \\ 1 \end{bmatrix}$ and expalin why this can be called the row averages.

6. The sales numbers of two companies in the first quarter are given below

	Jan	Feb	Mar
Company I	600	800	900
Company II	800	700	400

pdrom is 3 because 3 columns

Which of the following matrix operations will give the sales averages by company and which will give the sales averages by month?

multiply add up add 1 so #s

(A) $\dfrac{1}{3}\begin{bmatrix} 600 & 800 & 900 \\ 800 & 700 & 400 \end{bmatrix}\begin{bmatrix} 1 \\ 1 \\ 1 \end{bmatrix}$

(B) $\dfrac{1}{3}\begin{bmatrix} 1 & 1 \end{bmatrix}\begin{bmatrix} 600 & 800 & 900 \\ 800 & 700 & 400 \end{bmatrix}$

(C) $\dfrac{1}{2}\begin{bmatrix} 1 & 1 \end{bmatrix}\begin{bmatrix} 600 & 800 & 900 \\ 800 & 700 & 400 \end{bmatrix}$ *Monthly Average*

(D) $\dfrac{1}{2}\begin{bmatrix} 600 & 800 & 900 \\ 800 & 700 & 400 \end{bmatrix}\begin{bmatrix} 1 \\ 1 \\ 1 \end{bmatrix}$

(E) $\dfrac{1}{3}\begin{bmatrix} 600 & 800 & 900 \\ 800 & 700 & 400 \end{bmatrix}$

(F) $\dfrac{1}{2}\begin{bmatrix} 600 & 800 & 900 \\ 800 & 700 & 400 \end{bmatrix}$

7. Which of the following is the 4 × 4 identity matrix?

(A) $\begin{bmatrix} 1 & 1 & 1 & 1 \\ 1 & 1 & 1 & 1 \\ 1 & 1 & 1 & 1 \\ 1 & 1 & 1 & 1 \end{bmatrix}$

(B) $\begin{bmatrix} 0 & 0 & 0 & 1 \\ 0 & 0 & 1 & 0 \\ 0 & 1 & 0 & 0 \\ 1 & 0 & 0 & 0 \end{bmatrix}$

(C) $\begin{bmatrix} 1 & 0 & 0 & 0 \\ 0 & 1 & 0 & 0 \\ 0 & 0 & 1 & 0 \\ 0 & 0 & 0 & 1 \end{bmatrix}$

(D) $\begin{bmatrix} 1 & 0 & 0 & 1 \\ 0 & 0 & 0 & 0 \\ 0 & 0 & 0 & 0 \\ 1 & 0 & 0 & 1 \end{bmatrix}$

8. If I is the 3 × 3 identity matrix, then

(a) $I\begin{bmatrix} 9 & 7 & 5 \\ 8 & 6 & 4 \\ 3 & 2 & 1 \end{bmatrix} = \begin{bmatrix} 9 & 7 & 5 \end{bmatrix}$ *Get Same Matrix*

(b) $I\begin{bmatrix} 5 \\ 1 \\ 2 \end{bmatrix} = \begin{bmatrix} 5 \\ 1 \\ 2 \end{bmatrix}$ *Same Matrix*

(c) $\begin{bmatrix} 5 \\ 1 \\ 2 \end{bmatrix} I = undefined$ *3×1 3×3*

(d) $\begin{bmatrix} a & b & c \\ x & y & z \end{bmatrix} I = Same\ matrix$ *2×3 3×3*

9. Suppose A and B are two 4 × 4 matrices. Determine whether each of the following statements is true or false.

Matrix · Inverse = Identity

(a) If $AB = I$, the 4 × 4 identity matrix, then A and B are inverses of each other. *True*

(b) If $BA = I$, then A and B are inverses of each other. *True*

(c) Since AB and BA are both defined, they must always equal each other, i.e., $AB = BA$. False

10. A matrix A and its inverse A^{-1} are given below. Solve the following systems (Do not use the Gauss-Jordan method here).

$$A = \begin{bmatrix} 1 & 2 & 1 \\ 2 & -1 & 3 \\ 2 & 2 & 1 \end{bmatrix} \qquad A^{-1} = \begin{bmatrix} -1 & 0 & 1 \\ 4/7 & -1/7 & -1/7 \\ 6/7 & 2/7 & -5/7 \end{bmatrix}$$

(a) $\begin{cases} x + 2y + z = -1 \\ 2x - y + 3z = 2 \\ 2x + 2y + z = 1 \end{cases}$

(b) $\begin{cases} x + 2y + z = 1 \\ 2x - y + 3z = 4 \\ 2x + 2y + z = 7 \end{cases}$

(c) $\begin{cases} -x + z = 5 \\ (4/7)x - (1/7)y - (1/7)z = 4 \\ (6/7)x + (2/7)y - (5/7)z = 3 \end{cases}$

(d) $\begin{cases} -x + z = -0.35 \\ (4/7)x - (1/7)y - (1/7)z = 0.44 \\ (6/7)x + (2/7)y - (5/7)z = -0.25 \end{cases}$

(e) $\begin{cases} x + 2y + z = 14 \\ 2x - y + 3z = 1 \\ 2x + 2y + z = -1 \end{cases}$

(f) $\begin{cases} x + 2y + z = 3 \\ 2x - y + 3z = 8 \\ 2x + 2y + z = 4 \end{cases}$

(g) $\begin{cases} -x + z = a \\ (4/7)x - (1/7)y - (1/7)z = b \\ (6/7)x + (2/7)y - (5/7)z = c \end{cases}$

11. Determine whether the following statement is true or false: The inverse of a square matrix, if it exists, can be found rather easily by existing software. True

Chapter 4

Counting

This chapter is about counting, not the "1, 2, 3, 4, 5, . . . " kind of counting—that's for lovelier creatures—but "in how many ways this can be done" kind of counting. For example, if there are 30 freshmen, 25 sophomores, and 20 juniors in a club, and a committee of 7 people is to be formed, in how many ways can the committee consist of 4 freshmen, 1 sophomore, and 2 juniors?

4.1 THE MULTIPLICATION PRINCIPLE

The *multiplication principle*, sometimes also called the *fundamental counting principle*, is the foundation of all counting methods.

> The Multiplication Principle (The Fundamental Counting Principle)
>
> If task A can be accomplished in m different ways, and task B can be accomplished in n different ways, then the sequence A followed by B can be accomplished in $m \times n$ different ways. $\hspace{1em}$ (4.1.1)

Example 4.1.1. *In how many ways can the figure – ○○ – be colored if we can color the small circle blue or green, and the large circle red, purple, or yellow?*

Solution. *Coloring the small circle is task A, which can be done in 2 different ways: blue or green; coloring the large circle is task B, which can be done in 3 different ways: red, purple, or yellow. So the total number of ways to color the figure is, according to (4.1.1), 2 × 3 = 6. Since 6 is small number, we can list all six color combinations just to be completely certain, and also gain a better understanding of the multiplication principle. Here are the six ways we can color the figure (we* *use the first letter of a color to denote the color, R for red, etc.): BR, BP, BY, GR, GP, GY◄*

Example 4.1.2. *There are three freeways connecting town L and town B, and four freeways connecting town B and town A. Suppose all freeways are trouble free and we can pick any freeway we want, in how many ways can we go from town L to town B to town A?*

Solution. *I know, too easy.* $3 \times 4 = 12$—*but maybe you want to list all 12 routes again* :) ◄

The multiplication principle can be naturally extended to three or more tasks.

Example 4.1.3. *I am going shopping at Costco (most exciting event of the week). I need one bag of popcorn, one case of soda, and one box of napkins. There are five different popcorns, seven different sodas, and three different napkins available. In how many ways can I complete my shopping order?*

Solution. $5 \times 7 \times 3 = 105$ *ways*◄

Example 4.1.4. *A license plate number consists of two letters chosen from the English alphabet (case insensitive), followed by three digits chosen from the set* $\{0, 1, 2, 3, 4, 5, 6, 7, 8, 9\}$. *How many different license plate numbers are possible?*

Solution. *There are* 26 *letters in the English alphabet. When we choose the first letter, we have* 26 *choices, and another* 26 *choices when we choose the second letter. After that we choose the three digits one at a time, and have* 10 *choices per digit. So the total number of ways we can form a license plate number is*

$$26 \times 26 \times 10 \times 10 \times 10 = 676000 ◄$$

|||| EXERCISES 4.1 ||||

1. You have three different puzzle books and four different story books. In how many ways can you take one of each to school for Show and Tell?

2. The president needs to assemble her team of Eight Imposing Figures. There are 3 candidates for the position of Secretary of Cyber Security, 4 candidates for the position of Secretary of Food Safety, 5 candidates for the position of Secretary of Border Crossability, 6 candidates for the position of Secretary of Vulnerability, 7 candidates for the position of Secretary of Presentability, 8 candidates for the Secretary of Bility, 9 candidates for the position of Secretary of Mobility, and 10 candidates for the Secretary of Visibility. In how many different ways can the team be assembled?

3. The English alphabet consists of 26 letters. The first letter of a person's first name and the first letter of this person's last name form the initials of this person. For example, the initials of John Smith are JS. How many different initials are possible in the English alphabet system?

4. How many different four-digit (a digit is a number chosen from the set $\{0, 1, 2, 3, 4, 5, 6, 7, 8, 9\}$) PIN's are there?

5. How many different four-digit PIN's are there if the first digit cannot be a 0?

6. A password consists of six characters. The first character must be a letter chosen from the English alphabet, each of the remaining five characters can be either a letter from the English alphabet or a digit from the set $\{0, 1, 2, 3, 4, 5, 6, 7, 8, 9\}$. How many different passwords are possible?

7. A razor-thin two-sided coin is flipped four times. How many different sequences of heads and tails are possible?

8. A Chinese Pie Stand offers a delicious triangular shaped, chewy, green onion pie. You can eat the pie by itself as the pie is very delicious as we just mentioned. If you are feeling up to it, you can also order the pie with any or all of four luxurious toppings: heavenly beef, earthly chicken, salty ocean fish, and spicy insanity crazy chili. In how many ways can a pie be ordered?

9. How many subsets does the set $\{w, x, y, z\}$ have?

10. The last three problems share the same mathematical structure. Make sure you see that.

11. Mr. McDonald's syllabus quiz consists of four True-or-False questions. In how many ways can the quiz questions be answered

    (a) If to each question a student must answer either True or False?

    (b) If to each question a student also has the choice of not answering?

4.2 PERMUTATIONS AND COMBINATIONS

Suppose there are 10 basketball teams in a conference. If you pick 3 teams and rank them as number one, number two, and number three, you are performing a *permutation* of 10 objects taken 3 at a time. The total number of ways this can be done is denoted by $P(10,3)$, $_{10}P_3$, P_3^{10}, or P_{10}^3. We will use the notation $P(10,3)$ here for its ease of typing.

$$P(10,3) = \text{number of permutations of 10 objects taken 3 at a time}$$

If you pick 3 teams out of 10, but do NOT RANK them from number one to number three, then you are performing a *combination* of 10 objects taken 3 at a time. The total number of ways this can be done is denoted by $C(10,3)$, $_{10}C_3$, C_3^{10}, C_{10}^3, or $\binom{10}{3}$. We will use the notation $C(10,3)$.

$$C(10,3) = \text{number of combination of 10 objects taken 3 at a time}$$

We have the notations and the definitions now. We will find the permutation and combination formulas next.

Let's use our ten-team conference to derive the formulas. There are two ways to choose the number one, number two, and number three teams from ten teams.

Method one: We choose one team as the number one team from ten teams, we have 10 choices when we do this. After that there are nine teams left and from these nine we choose one team as the number two team, we have 9 choices here. There are eight teams left after the number two team is chosen, and we choose one of them as the number three team, we have 8 choices. According to the multiplication principle, the total number of ways this process can be done is $10 \times 9 \times 8 = 720$. Remember this a permutation of 10 objects taken 3 at a time so

$$P(10,3) = 10 \cdot 9 \cdot 8$$

Notice that 10 is the first number, followed by 9, followed by 8. We stop at 8 because we are multiplying THREE numbers as indicated by the number 3 in $P(10,3)$.

Method two: Another way to choose the number one, number two, and number three teams from ten teams is to (a) first choose three teams from ten teams—a combination of 10 objects taken 3 at a time, and then (b) rank these three teams from number one to number three. Step (a) can be done in $C(10,3)$ ways and step (b) can be done in $P(3,3)$ ways. According to the multiplication principle, the total number of ways to perform steps (a) and (b) as a sequence is $C(10,3)P(3,3)$.

Since method one and method two are equivalent, we can be sure that

$$P(10,3) = C(10,3)P(3,3)$$

or

$$C(10,3) = \frac{P(10,3)}{P(3,3)} = \frac{10 \cdot 9 \cdot 8}{3 \cdot 2 \cdot 1}$$

Listed below are the general formulas, where n and k are nonnegative integers with $k \le n$.

Permutation of n objects taken k at a time

$$P(n,k) = \overbrace{n(n-1)\cdots(n-k+1)}^{k \text{ consecutive integers from } n \text{ to } (n-k+1)}$$

$$(4.2.1)$$

Combination of n objects taken k at a time

$$C(n,k) = \frac{\overbrace{n(n-1)\cdots(n-k+1)}^{k \text{ consecutive integers from } n \text{ to } (n-k+1)}}{\underbrace{k(k-1)\cdots2\cdot1}_{k \text{ consecutive integers from } k \text{ to } 1}}$$

dont work w/ a 0

$$(4.2.2)$$

Example 4.2.1. *Let's compute some permutations and combinations.*

(a) $P(5,3) = 5 \cdot 4 \cdot 3 = 60$

(b) $C(5,3) = \dfrac{5 \cdot 4 \cdot 3}{3 \cdot 2 \cdot 1} = 10$

(c) $P(5,5) = 5 \cdot 4 \cdot 3 \cdot 2 \cdot 1 = 120$

(d) $C(5,5) = \dfrac{5 \cdot 4 \cdot 3 \cdot 2 \cdot 1}{5 \cdot 4 \cdot 3 \cdot 2 \cdot 1} = 1$

(e) $P(5,1) = 5$

(f) $C(5,1) = \dfrac{5}{1} = 5$

(g) $P(5,0) = ?$ *We are unable to use formula* (4.2.1) *here because there is no way to multiply* 0 *consecutive integers together. Is* $P(5,0)$ *undefined, then? Well, it is actually defined, and we will soon see how it is computed* ◄

Formulas (4.2.1) and (4.2.2) are good when we perform the computations by hand, and when $k \geq 1$. The formulas have two drawbacks: (a) they look cumbersome with the dots in them; and (b) they don't work when $k = 0$.

We will utilize the *factorial* notation to rewrite (4.2.1) and (4.2.2) in more compact forms, and use the new forms to remedy the two drawbacks at the same time.

Let n be a positive integer. The product $n \cdot (n-1) \cdots 2 \cdot 1$ is called "n factorial", and is denoted by $n!$.

---

n factorial for positive integers $n \geq 1$

$$n! = n \cdot (n-1) \cdots 2 \cdot 1$$

$$(4.2.3)$$

0 factorial is defined to be 1, $0! = 1$. The definition is not arbitrary, though. Intuitively, we can reason that since $P(3, 3) = 3!$ is the number of ways we can arrange 3 objects in order, $P(0, 0) = 0!$ must then be the number of ways we can arrange 0 objects in order, and the number is 1 because there are no objects at all, leaving us with only ONE option: "do nothing".

For the math inclined, the definition $0! = 1$ can be explained by the general definition of factorial. The general (when n doesn't equal a positive integer) factorial is defined by the Gamma function: $n! = \Gamma(n+1) = \int_0^\infty x^n e^{-x} dx$. Those who know how to perform the integration can verify that this definition agrees with (4.2.3) for $n = 1, 2, 3, \cdots$. This function can then be used to define the factorial of a non-integer such as 3.5!, for example. And if we plug in 0 for n, the function gives us $0! = \Gamma(1) = 1$. If you feel a little lost here, don't worry about this little paragraph and just move on. If you have never seen this before but have taken calculus, you can do some investigation to find out more about the Gamma function.

Zero factorial

$$0! = 1 \tag{4.2.4}$$

We can now rewrite the permutation and combination formulas using the factorial notation.

Example 4.2.2. *Write a permutation and a combination using the factorial notation.*

$$P(5, 3) = 5 \cdot 4 \cdot 3 = \frac{5 \cdot 4 \cdot 3 \cdot 2 \cdot 1}{2 \cdot 1} = \frac{5!}{(5-3)!}$$

$$C(5, 3) = \frac{5 \cdot 4 \cdot 3}{3 \cdot 2 \cdot 1} = \frac{5 \cdot 4 \cdot 3 \cdot 2 \cdot 1}{3 \cdot 2 \cdot 1 \cdot 2 \cdot 1} = \frac{5!}{3!(5-3)!} \blacktriangleleft$$

The general formulas are

Permutation of n objects taken k at a time

$$P(n, k) = \frac{n!}{(n-k)!} \tag{4.2.5}$$

Combination of n objects taken k at a time

$$C(n, k) = \frac{n!}{k!(n-k)!} \tag{4.2.6}$$

With these formulas we can compute some permutations and combinations that we couldn't before.

Example 4.2.3. *Use (4.2.5) and (4.2.6) to compute some permutations and combinations.*

$$P(9,4) = \frac{9!}{(9-4)!} = \frac{9!}{5!} = \frac{9 \cdot 8 \cdot 7 \cdot 6 \cdot 5 \cdot 4 \cdot 3 \cdot 2 \cdot 1}{5 \cdot 4 \cdot 3 \cdot 2 \cdot 1} = 3024$$

$$C(9,4) = \frac{9!}{4!(9-4)!} = \frac{9!}{4!5!} = \frac{9 \cdot 8 \cdot 7 \cdot 6 \cdot 5 \cdot 4 \cdot 3 \cdot 2 \cdot 1}{(4 \cdot 3 \cdot 2 \cdot 1) \cdot (5 \cdot 4 \cdot 3 \cdot 2 \cdot 1)} = 126$$

$$P(5,0) = \frac{5!}{(5-0)!} = \frac{5!}{5!} = 1 \qquad \text{\textit{we couldn't do this using (4.2.1)}}$$

$$C(5,0) = \frac{5!}{0!(5-0)!} = \frac{5!}{1 \cdot 5!} = 1 \qquad \text{\textit{we couldn't do this using (4.2.2)}} \blacktriangleleft$$

The fact that $P(5,0) = 1$ is not too surprising. $P(5,0)$ is the permutation of 5 objects taken 0 at a time. If no object is taken, then "do nothing" is the ONLY way to do it, hence the answer of 1.

How do we differentiate a permutation from a combination? A permutation always has some underlying order or ranking involved, whereas a combination does not. Study the next example carefully.

Example 4.2.4. *Suppose three scholarships are to be awarded to three students from a pool of 20 applicants. Find the total number of ways the scholarships can be awarded if*

   (a) The scholarships are in the amounts of $12000, $8000, and $5000.
   (b) The scholarships are in the amounts of $8000, $8000, and $8000.

Solution.

   (a) This is a permutation even though the word "order" or "rank" is not explicitly used. The three scholarships are different, and it needs to be decided who receives $12000, who receives $8000, and who receives $5000. The answer is

$$P(20,3) = 6840$$

   (b) This is a combination because all three scholarships are identical, and there is no ranking involved. The answer is

$$C(20,3) = 1140 \blacktriangleleft$$

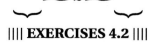

|||| **EXERCISES 4.2** ||||

   1. Perform the following operations.

(a) 6!	(g) 0!	(m) $P(6,1)$
(b) 5!	(h) $P(6,6)$	(n) $P(6,0)$
(c) 4!	(i) $P(6,5)$	(o) $C(6,6)$ and $C(6,0)$
(d) 3!	(j) $P(6,4)$	(p) $C(6,5)$ and $C(6,1)$
(e) 2!	(k) $P(6,3)$	(q) $C(6,4)$ and $C(6,2)$
(f) 1!	(l) $P(6,2)$	(r) $C(6,3)$

2. Explain, without doing any computations, why $C(1000000,999998) = C(1000000,2)$.

3. The Math Department Textbook Selection Committee is considering adopting a new finite math textbook. In how many ways can the committee select three books from a list of ten books for further evaluation?

4. In how many ways can the Starving Poets' Society select the number one, number two, and number three poems of the year from ten submissions?

5. Mrs. Smith's third grade class has 12 students. In how many ways can Mrs. Smith choose two students to assist her in a class demonstration?

6. In how many ways can Mrs. Smith give a chocolate bar, a bag of potato chips, and a can of cane juice to three of the twelve students in her class, one item per student?

7. The Rocket Man built a spaceship in his backyard. In addition to the driver (which by default will be the Rocket Man himself, of course), the spaceship can hold four passengers. The Rocket Man has nine loyal friends. In how many ways can the Rocket Man pick four of his nine loyal friends to join him on the maiden voyage?

8. How many subsets does a set with four elements have? This problem is also in the last section, but this time find the answer by finding the number of subsets that have zero elements, one element, two elements, three elements, and four elements, and then adding the results together.

4.3 MIXED COUNTING PROBLEMS

Counting problems can become complex, and they are in one class all by themselves. We will only touch the surface here, because (1) this is a finite mathematics text and counting is only a part of it, and (2) the author of this text doesn't know too much about counting.

We have the basic ingredients: the multiplication principle and the combination and permutation formulas. We will use these to investigate some interesting problems.

First let's visit an old friend, formula (2.5.1), the inclusion-exclusion formula for sets

$$n(A \cup B) = n(A) + n(B) - n(A \cap B)$$

If the two sets A and B are *mutually exclusive*, i.e., they do not have any common elements, then $A \cap B = \emptyset$, hence $n(A \cap B) = 0$, and the formula becomes

---

Inclusion-exclusion formula for mutually exclusive sets

If $A \cap B = \emptyset$, then

$$n(A \cup B) = n(A) + n(B) \tag{4.3.1}$$

Often sets are given by verbal descriptions and their elements are not explicitly given. In such cases how do we determine if two sets are mutually exclusive? An easy way to determine if two sets are mutually exclusive is to check that "if A happens, then B cannot happen, and vice versa."

We mention this because sometimes we need to count the number of elements in a set that contains several mutually exclusive subsets. We need to be able to distinguish between the multiplication principle, which involves multiple steps in one task, and formula (4.3.1), which is used when a task consists of several mutually exclusive sub-tasks.

Example 4.3.1. *A two sided coin is flipped five times and the sequence of heads and tails is recorded.*

(a) *How many different sequences are possible?* Permutation

(b) *How many different sequences contain exactly 2 heads and 3 tails?* Combinations

(c) *How many different sequences contain exactly 3 heads and 2 tails?* Combination

(d) *How many different sequences contain exactly one heads (and four tails, of course)?*

(e) *How many different sequences contain at least one heads?*

Solution.

(a) *Every time we flip a coin, there are two possible outcomes: heads or tails. So if we flip a coin five times, the total number of heads-tails sequences is $2^5 = 32$.*

(b) *Let's use H to represent heads and T tails. Some possible sequences that contain 2 heads and 3 tails are: $HHTTT, HTHTT, THHTT$. We don't want to list all the possible sequences and then count them because that is too daunting a task if we flip a coin 100 times. We look at the three sequences we listed here, and we see that there are always two H's and three T's, but their positions are not fixed. We can put two H's in any two of the five spots and put three T's in the remaining three spots. Now the question becomes "in how many ways can we choose two spots out of five to put H's in?" Because once the two spots for two H's are chosen,* *the remaining three spots are automatically for three T's. According to the multiplication principle, the final answer is $C(5,2)C(3,3) = C(5,2) = 10$.*

(c) *This is an easy one after the last one. The answer is $C(5,3) = 10$. Compare this to the answer above and we see $C(5,2) = C(5,3)$—more on this later.*

(d) *$C(5,1) = 5$.*

(e) *"At least one heads" means there could be "exactly one heads", or "exactly two heads", or "exactly three heads", or "exactly four heads", or "exactly five heads". And notice that all these possible outcomes are mutually exclusive because if a sequence contains exactly one heads, then it is impossible for that same sequence to contain exactly two heads, and so on. And so according to formula (4.3.1), the answer is*

$$C(5,1) + C(5,2) + C(5,3) + C(5,4) + C(5,5) = 31$$

31 is of course the correct answer. But if we flip a coin 50 times and we ask the same question, we can imagine what kind of trouble we will get ourselves into if we try to do it the same way: we will have to count all the sequences that contain one heads, two heads, three heads, ... all the way to fifty heads. Well, sometimes we have to do things the hard way, and we will put our head down and do it. But here in this particular situation, there is a better way! (This is the exclamation mark, it is not "way factorial".) Notice that when a coin is flipped five times, there are six possible numbers of heads in a sequence: $0, 1, 2, 3, 4, 5$. When we ask for "at least one heads" we are asking for the sequences that contain 1, 2, 3, 4, or 5 heads. We already know the total number of possible sequences is 32 from the first question of this example, so here we can simply get the answer by subtracting the number of sequences that contain zero heads from the total: $32 - C(5,0) = 32 - 1 = 31$ ◄

The method we just used above is actually formula (4.3.1). If U is the universal set, and E is a subset of U, then E and E' are mutually exclusive and $U = E \cup E'$. According to (4.3.1), $n(U) = n(E) + n(E')$, so $n(E) = n(U) - n(E')$. Use the example above, $U = \{$all sequences$\}$, $E = \{$sequences with at least one heads$\}$, and $E' = \{$sequences with zero heads$\}$. $n(E) = n(U) - n(E') = 32 - 1 = 31$.

Example 4.3.2. *A three-sided die has the numbers 1, 2, and 3 on its three faces, one on each face. Roll the die seven times and record the sequence of numbers observed.*

(a) *How many different sequences are possible?*

(b) *How many sequences contain four 1's, two 2's, and one 3?*

Solution. *This is a generalization of the previous example—every step has three possible outcomes instead of two. Here is a possible sequence:* **1123121**.

(a) *Every roll has three possible outcomes, the total number of sequences is $3^7 = 2187$.*

(b) *The four 1's occupy 4 of the 7 spots, the number of ways this can occur is $C(7,4)$. After that the* *two 2's will occupy 2 of the remaining 3 spots, this can occur in $C(3,2)$ ways. The final spot is for the one 3. The total number of ways this sequence can occur is $C(7,4)C(3,2)C(1,1) = 35 \cdot 3 \cdot 1 = 105$* ◄

The examples have been very wordy, but that's necessary in the beginning when every little step needs to be explained. Things will get better gradually.

Example 4.3.3. *There are six red cards and five green cards in a box. The cards are numbered so any one card can be distinguished from another card of the same color. A sample of four cards is taken from the box.*

- *(a) How many different samples are possible?*
- *(b) How many samples contain one green and three red cards?* multiply
- *(c) How many samples contain at least one red card?*

Solution.

- *(a) We are taking four objects out of eleven objects, and order is not important because we are only paying attention to colors. Answer: $C(11,4) = 330$.*
- *(b) To have one green and three red cards in a sample, we can take any three red cards from the six red cards, and take any green card from the five green cards. This is one task that involves two steps so by the multiplication principle, the number of samples that contain one green and three red cards is $C(6,3)C(5,1) = 100$.*
- *(c) If we count the numbers of red and green cards in the sample, the sample space is $U = \{0R4G, 1R3G, 2R2G, 3R1G, 4R0G\}$. Let $E = \{$at least one red card$\} = \{1R3G, 2R2G, 3R1G, 4R0G\}$, then $E' = \{$no red cards$\} = \{0R4G\}$. $n(E') = C(6,0)C(5,4) = 5$. So the answer is $n(E) = n(U) - n(E') = 330 - 5 = 325$ samples that contain at least one red card* ◄

Example 4.3.4. *There are five red, four yellow, and three green numbered cards in a box. A sample of six cards is taken from the box.*

- *(a) How many different samples are possible?*
- *(b) How many samples contain three red, two yellow, and one green card?* multiply

Solution. *This is a generalization of the previous example.*

- *(a) Twelve cards, we take six, number of ways $= C(12,6) = 924$.*

(b) Take threes cards from the five red cards: $C(5,3)$ ways; take two cards from the four yellow cards: $C(4,2)$ ways; take one card from the three green cards: $C(3,1)$ ways. Final answer: $C(5,3)C(4,2)C(3,1) = 180$ samples◀

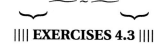

|||| **EXERCISES 4.3** ||||

1. A password consists of six characters. The first character must be a letter chosen from the English alphabet, one of the remaining five characters must be a digit chosen from the set $\{0,1,2,3,4,5,6,7,8,9\}$, and each of the remaining four characters must be a special character from the set $\{!,@,\#,\$,\%,\&\}$. How many different passwords are possible?

2. Find the number of ways three different math books, four different English books and five different biology books can be arranged in a row on a shelf

    (a) If the books can be arranged in any order.

    (b) If the math books must be next to each other.

    (c) If the English books must be next to each other.

    (d) If the math books must be next to each other and the biology books must be next to each other.

    (e) If the books of each subject must be next to each other.

    (f) If the books of each subject must be next to each other and the subjects must be in the order of math-English-biology.

3. A coin is flipped six times and the sequence of heads and tails is recorded. How many different sequences consisting of exactly two heads and four tails are possible?

4. A coin is flipped seven times and the sequence of heads and tails is recorded. How many different sequences consisting of more heads than tails are possible?

5. President Prezident is surrounded by twelve trusted bodyguards all the time. Five of the bodyguards are heavily armed, the other seven are lightly legged. On a trip to a fourth galaxy planet to negotiate with a dictator named Dictater, Prezident is threatened by a wild dragon of the third kind. In how many ways can Prezident deploy two heavily armed and three lightly legged bodyguards to subdue the dragon?

6. Emma has eighteen different teddy bears. In how many ways can Emma divide the bears into three teams, one with five bears, one with six, and one with seven, and further name a captain for each team?

NO combination to solve

7. Emma has eighteen different teddy bears, one of them is her favorite. She names her favorite bear the President, and then picks five bears to form the Senate and names one of the five as the Senate Chair. The remaining twelve will be the citizens of the Democratic Republic of Bears. In how many ways can Emma accomplish this bear nation building task?

8. Four members from the math department, four members from the biology department, and four members from the English department have declared themselves candidates for two vacant faculty senate seats. Suppose the two seats cannot be filled by two members from the same department. In how many ways can the two seats be filled?

9. The PAC-12 conference, as its name suggests, has twelve member schools.

    (a) If during a regular season the University of Arizona's football team has to play every other school in the conference exactly once, how many games does the U of A team play?

    (b) If during a regular season each of the 12 teams in the conference has to play every other team in the conference exactly once, how many conference games will be played?

    (c) Suppose in the conference there are six pairs of rivalries (the University of Arizona and Arizona State University are a pair, for example). If during a regular season each of the 12 teams in the conference has to play every other non-rivalry team in the conference exactly once, but the rivalry team twice, how many conference games will be played?

4.4 PARTITION—A UNIFIED APPROACH TO PERMUTATIONS AND COMBINATIONS

Albert Einstein devoted a large portion of his life to a Unified Field Theory. The idea is simple yet ambitious: If one stone can kill two birds, why use two stones? Better yet, if one stone can kill all the birds, why use all the stones?

We are of course not as ambitious as, and far less talented than, Albert. But we can still put permutations and combinations under one umbrella called *partition*.

Recall that from example 4.3.1, $C(5,2) = C(5,3)$. This is not a mere coincidence. In general, $C(n,k) = C(n, n-k)$. The formula is easy to verify using formula (4.2.6), and the reader is encouraged to do so. We want to point out the intuitive part here, though, because intuition and imagination are very important in innovation as long as people remember to verify them. So why is $C(5,2) = C(5,3)$ intuitive? $C(5,2)$ is the number of ways we can choose 2 objects out of 5 without putting them in order, it is therefore the operation of dividing 5 objects into two groups, one with 2 objects and the other with 3. But his operation is exactly the same as the operation of dividing 5 objects into a 3-object group and a 2-object group. The second operation can be done in $C(5,3)$ ways, therefore $C(5,2) = C(5,3)$.

Here is one more example: If you are giving two prizes to two people among 100 people, you can say to the two people who win the prizes "hey, you two guys win the prizes," or you can say to the 98 people who don't win the prizes "hey, you 98 guys don't win no nothing." The first scenario is a $C(100,2)$ and the second scenario is a $C(100,98)$, and they accomplish the exact same task, hence the identity $C(100,2) = C(100,98)$.

So $C(n,k)$ is the number of ways to divide n objects into two groups, one with k objects and the other $(n-k)$ objects. The natural generalization is to divide n objects into three or more groups.

Example 4.4.1. *OverNiteRiches hauled 30 people from Tucson, Arizona to Las Vegas, Nevada for a once-in-a-lifetime experience. Hotel rooms are hard to find though and the 30 people had to be split into four groups: 15 will stay at hotel A, 10 will stay at hotel B, 3 will stay at hotel C, and 2 will stay at hotel D. In how many ways can the 30 people be arranged to stay at the four hotels?*

Solution. *Imagine you are the tour guide. First you pick 15 out of 30 people and send them to hotel A, there are $C(30,15)$ ways to do that. After the first 15 people are picked, there are 15 people left and you pick 10 to go to hotel B, there are $C(15,10)$ ways to do it. Now there are 5 people left and you pick 3 to enjoy hotel C, there are $C(5,3)$ ways to accomplish this. After that there are two people left and you drive them to hotel D, 2 people choose 2, $C(2,2)$ ways to do it. This entire process involves four steps and by the multiplication principle, the total number of ways this can occur is $C(30,15)C(15,10)C(5,3)C(2,2)$. We will not go straight to the final numerical answer, we will use formula (4.2.6) to rewrite our answer and obtain a general formula.*

$$C(30,15)C(15,10)C(5,3)C(2,2)$$
$$= \frac{30!}{15!15!}\frac{15!}{10!5!}\frac{5!}{3!2!}\frac{2!}{2!0!}$$
$$= \frac{30!}{15!10!3!2!}$$

Stare at the result, stare at the result... what a beauty! 30 objects divided into four groups, the first group has 15 objects, the second 10, the third 3, and the fourth 2. You see the answer has 30! in the numerator, then 15!, 10!, 3!, and 2! in the denominator—that is unbelievable elegance◄

What has just happened above is called a *partition*. The official definition is a little wordy so we will leave it to the curious people to find out the exact language. Here is what it is: We take 30 distinct objects, divide them into 4 distinct groups of 15, 10, 3 and 2 objects. The number of ways this partition can be accomplished is written as

$$\binom{30}{15,10,3,2}$$

In general, if we divide n distinct objects into m distinct groups of n_1, n_2, ..., n_m objects, then the number of ways this can be accomplished is

---

Partition of n distinct objects into m distinct groups

$$\binom{n}{n_1, n_2, \ldots, n_m} = \frac{n!}{n_1! n_2! \cdots n_m!} \tag{4.4.1}$$

where $n_1 + n_2 + \cdots + n_m = n$

Example 4.4.2. *Forty-five math teachers went to a conference they didn't care about but went to anyway because free lunch was provided. There were 15 hamburger lunches, 10 turkey sandwich lunches, 10 fried fish lunches, 6 egg and cheese lunches, and 4 vegan lunches. In how many ways can the lunches be distributed, one for each math teacher?*

Solution.

$$\binom{45}{15,10,10,6,4} = \frac{45!}{15!10!10!6!4!} \approx 4 \times 10^{26} \blacktriangleleft$$

What is the connection between partition and permutation and combination? Obviously,

$$C(n,k) = \binom{n}{k, n-k}$$

and this easily explains why $C(n, k) = C(n, n - k)$.

$P(n, k)$ can be viewed as the number of ways to divide n objects into $(k + 1)$ groups, the first k groups each has 1 object, and the $(k + 1)^{\text{st}}$ group has $(n - k)$ objects:

$$P(n,k) = \frac{n!}{(n-k)!} = \frac{n!}{\underbrace{1! \cdots 1!}_{k \ 1!s} \cdot (n-k)!} = \binom{n}{\underbrace{1, \ldots, 1}_{k \ 1s}, n-k}$$

We have thus achieved a mini-unification—we have put both permutation and combination under the broader umbrella of partition.

Example 4.4.3. *The Occupy-All-Streets movement has gathered 20 members. They decide to elect among themselves a president, two vice presidents, and five propaganda managers. How many different election outcomes are possible?*

[handwritten: Have to add up to total to make partition]

Solution. *Method 1: We use combinations and the multiplication principle to find the answer. There are also variations within this method and they really show what is quite unique in counting, and that is sometimes we can know for sure something has to be true without having to verify it mathematically. We show two variations below. The reader is encouraged to discover a few more.*

(a) *We choose a president first, there are $C(20,1)$ ways; then choose two vice presidents, $C(19,2)$ ways (19 people available for VP after one person has been chosen as the president); then choose five propaganda managers, $C(17,5)$ ways. Answer: $C(20,1)C(19,2)C(17,5) = 21162960$.*

(b) *We first choose the 8 officials from among the 20 members, $C(20,8)$ ways; after that we choose a president from these 8 people, $C(8,1)$ ways; then two VP's, $C(7,2)$ ways; the remaining five are automatically propaganda managers. Answer: $C(20,8)C(8,1)C(7,2) = 21162960$.*

Method 2: We use partition. 20 *people are divided into four groups:* 1 *president,* 2 *vice presidents,* 5 *propaganda managers, and* 12 *members. Answer:*

$$\binom{20}{1,2,5,12} = \frac{20!}{1!2!5!12!} = 21162960 \blacktriangleleft$$

All roads lead to Rome—all the expressions below equal 21162960. Can you see the reasoning behind each expression?

~~~ **|||| EXERCISES 4.4 ||||**

1. Fill in the blank to complete the formula.

(a) $C(10,3) = \dfrac{10!}{3!\underline{}}$ (b) $P(10,3) = \dfrac{10!}{1!1!1!\underline{}}$ (c) $C(10,3) = \dfrac{P(10,3)}{\underline{}}$

2. The scout leader leads fourteen scouts to this remote forest. To settle for the evening, five people need to go hunting, three people need to find water, four people need to collect firewood, and two people need to set up tents. In how many different ways can the leader assign the jobs to the fourteen scouts?

3. In how many ways can one $12,000 scholarship, three $8,000 scholarships, and six $5,000 scholarships be awarded to ten students, one scholarship for each student, out of eighteen applicants?

4. Emma has eighteen different teddy bears. In how many ways can Emma give five bears red hats, six white hats, and seven blue hats?

5. Emma has eighteen different teddy bears. In how many ways can she name one bear the President, one the Vice President, one the Speaker of the Den, and five Senior Advisors to the President?

6. A home builder builds fifteen identical homes along a beach. In how many ways can three of the homes be painted red, five painted yellow, and seven painted green?

Chapter 5

From Counting to Probability

If I flip a fair coin once, I may get heads or tails. If I flip the same coin many, many times, then I will get some heads and some tails. As the number of flips increases, the ratio: (number of heads) to (total number of flips) will approach 1 to 2. We call $1/2 = 0.50$ the probability of getting heads when a fair coin is flipped. We will define probability later.

5.1 TERMINOLOGIES AND NOTATIONS

1. An activity that can be repeated under identical conditions is called an *experiment*.
2. Possible results of an experiment are called *outcomes*.
3. The set of all possible outcomes of an experiment is called the *sample space* (this is the universal set in chapter 2).
4. Any subset of the sample space is called an *event*.

Example 5.1.1. *Flip a coin three times and observe the sequence of heads and tails.*

  (a) Find the sample space.
  (b) Find all the outcomes in the event "there are two heads and one tails".
  (c) Find all the outcomes in the event "there is at least one heads".

Solution. *We use H to represent the outcome "heads" and T "tails".*

  (a) $S = \{HHH, HHT, HTH, HTT, THH, THT, TTH, TTT\}$
  (b) $\{Two\ heads\ and\ one\ tails\} = \{HHT, HTH, THH\}$
  (c) $\{At\ least\ one\ heads\} = \{HHH, HHT, HTH, HTT, THH, THT, TTH\}$ ◄

Depending on what we are observing, the same physical experiment can produce different sample spaces. Compare the next example with the last one.

Example 5.1.2. *Flip a coin three times and record the number of heads in the sequence. Find the sample space.*

Solution. *The physical experiment is the same as the one in the last example, but what we keep track of is different. Here we are counting the number of heads in a sequence. A sequence can contain zero, one, two, or three heads. The sample space is*

$$S = \{0, 1, 2, 3\} \blacktriangleleft$$

The table below shows the connection between the two sample spaces in examples 5.1.1 and 5.1.2.

Example 5.1.2 outcomes	Example 5.1.1 outcomes
0	TTT
1	HTT, THT, TTH
2	HHT, HTH, THH
3	HHH

The *probability* of an event measures how likely the event will occur when an experiment is performed. To make this measurement uniform, we agree that a probability should be a real number ranging from 0 to 1. If the probability of an event is 0, we say the event is an *impossible event*. If the probability of an event is 1, we say the event is a *certain event*. If the probability of an event is 0.30, that means the ratio (number of times the event occurs) to (total number of experiments performed) will approach 30 to 100 as the number of experiments approaches infinity. But be careful here though, this does not mean if we perform the experiment 100 times, the event will occur exactly 30 times.

Given that the sample space of an experiment is $S = \{c_1, c_2, \ldots, c_k\}$.

1. We denote by $\Pr(c_i)$ the probability of outcome c_i, $i = 1, 2, \ldots k$.
2. We say an event occurs if the experiment results in an outcome that is contained in the event. For example, if $E = \{c_2, c_3, c_4\}$, then E occurs if the result of the experiment is either c_2, c_3, or c_4.
3. The probability of an event equals the sum of probabilities of all outcomes in the event. For example, if $E = \{c_2, c_3, c_4\}$, then $\Pr(E) = \Pr(c_2) + \Pr(c_3) + \Pr(c_4)$.
4. $0 \le \Pr(E) \le 1$ for any event E.
5. Since the sample space contains all the possible outcomes, $\Pr(S) = \Pr(c_1) + \Pr(c_2) + \cdots + \Pr(c_k) = 1$. In other words, the sample space is a certain event because every time the experiment is performed, the result will be one of the outcomes in the sample space.
6. $\Pr(\emptyset) = 0$, i.e., the empty set is an impossible event.

:: **EXERCISES 5.1** ::

1. A fair six-sided die has the numbers 1, 2, 3, 4, 5, 6 on its six faces, one on each face. The die is rolled once and the number on the side facing up after the die comes to a rest is recorded.

    (a) Find the sample space of the experiment.

    (b) Find the event {an even number is rolled}.

    (c) Find the event {a number greater than 2 is rolled}.

2. A coin is flipped three times and the sequence of heads and tails is recorded.

    (a) Find the sample space of the experiment.

    (b) Find the event {there are more heads than tails in the sequence}.

    (c) Find the event {the second flip results in heads}.

3. If the sample space of an experiment is $S = \{a, b, c\}$, then $\Pr(a) + \Pr(b) + \Pr(c) = $ ____ .

4. If the sample space of an experiment is S, then $\Pr(S) = $ ____ .

5. $\Pr(\emptyset) = $ ____ .

6. Can the probability of an event be 1.01?

7. Can the probability of an event be –0.5?

8. Can the probability of an event be greater than 1 or less than 0?

9. If the sample space of an experiment is $S = \{a, b, c\}$, and $\Pr(a) = 0.38$, $\Pr(c) = 0.52$, then $\Pr(b) = $ ____ .

10. Suppose the sample space of an experiment is $S = \{a, b, c, d, e, f\}$ with $\Pr(a) = 0.05$, $\Pr(b) = 0.15$, $\Pr(c) = 0.25$, $\Pr(d) = 0.35$, $\Pr(e) = 0.12$, $\Pr(f) = 0.08$. Let E be the event "a vowel is chosen from S". Find $\Pr(E)$.

5.2   EXPERIMENTS WITH EQUALLY LIKELY OUTCOMES

Example 5.2.1. *Flip a fair coin three times.*

(a) *Find the probability that there is exactly one heads in the sequence.*
(b) *Find the probability that there are exactly two heads in the sequence.*
(c) *Find the probability that there is at least one heads in the sequence.*
(d) *Find the probability that the first flip results in heads.*
(e) *Find the probability that the first flip results in heads AND there are exactly two heads in the sequence.*
(f) *Find the probability that the first flip results in heads OR there are exactly two heads in the sequence.*

Solution. *We have seen this experiment in the last section. If we actually record the sequence, the sample space is*

$$\{HHH, HHT, HTH, HTT, THH, THT, TTH, TTT\}$$

Since the coin is fair, each of these 8 sequences is equally likely to occur, and the sum of the probabilities of all the outcomes in the sample space must equal 1, the probability of each outcome must be 1/8:

$$\Pr(HHH) = \Pr(HHT) = \cdots = \Pr(TTT) = \frac{1}{8}$$

(a) *If we let E be the event {there is one heads in the sequence}, then*

$$E = \{HTT, THT, TTH\}$$

$$\Pr(E) = \frac{1}{8} + \frac{1}{8} + \frac{1}{8} = \frac{3}{8}$$

(b) *If we let F be the event {there are two heads in the sequence}, then*

$$F = \{HHT, HTH, THH\}$$

$$\Pr(F) = \frac{1}{8} + \frac{1}{8} + \frac{1}{8} = \frac{3}{8}$$

(c) *If we let G be the event {there is at least one heads in the sequence}, then*

$$G = \{HHH, HHT, HTH, HTT, THH, THT, TTH\}$$

$$\Pr(G) = \frac{1}{8} + \frac{1}{8} + \frac{1}{8} + \frac{1}{8} + \frac{1}{8} + \frac{1}{8} + \frac{1}{8} = \frac{7}{8}$$

If you are thinking "is there a better way to solve this problem?" congratulations! Your brain has been rewired. Instead of counting the probabilities of 7 outcomes, we can count the probability of the one outcome that is not in the event, then subtract this probability from 1

(1 being the probability of the sample space) to get the same answer:

$$\Pr(G) = 1 - \Pr(G') = 1 - \Pr(\{TTT\}) = 1 - \frac{1}{8} = \frac{7}{8}$$

(d) Let K be the event in question.

$$K = \{HHH, HHT, HTH, HTT\}$$

$$\Pr(K) = \frac{1}{8} + \frac{1}{8} + \frac{1}{8} + \frac{1}{8} = \frac{4}{8} = \frac{1}{2}$$

(e) Remember AND means INTERSECTION (section 2.3). So we are looking for $\Pr(K \cap F)$*:*

$$K \cap F = \{HHT, HTH\}$$

$$\Pr(K \cap F) = \frac{1}{8} + \frac{1}{8} = \frac{2}{8} = \frac{1}{4}$$

(f) OR stands for UNION (section 2.3). So we are looking for $\Pr(K \cup F)$*:*

$$K \cup F = \{HHH, HHT, HTH, HTT, THH\}$$

$$\Pr(K \cup F) = \frac{1}{8} + \frac{1}{8} + \frac{1}{8} + \frac{1}{8} + \frac{1}{8} = \frac{5}{8} \blacktriangleleft$$

This example lends us a few observations.

1. If an experiment has *equally likely outcomes*, then

> **Probability of an event in an experiment with equally likely outcomes**
>
> $$\Pr(E) = \frac{\text{number of outcomes in event } E}{\text{number of outcomes in the sample space}} \qquad (5.2.1)$$

   We can see this from the example above, there are 3 outcomes in event E, 8 outcomes in the sample space. Since all outcomes are equally likely, $\Pr(E) = 3/8$.

2. For any event E,

> **Probability of the complement event**
>
> $$\Pr(E) = 1 - \Pr(E') \qquad (5.2.2)$$

This formula, however, is not restricted to experiments with equally likely outcomes. It is true for any experiment.

3. From the example above we can verify that

> **Inclusion-exclusion formula for probabilities**
>
> $$\Pr(K \cup F) = \Pr(K) + \Pr(F) - \Pr(K \cap F) \qquad (5.2.3)$$

This formula is again true for any experiment. If it looks familiar to you, that's because it has the exact same structure as the inclusion-exclusion formula for sets in section 2.5◀

Example 5.2.2. *Fifteen dogs and twenty cats live in a gated community. They want to randomly form a Home Owners Association (HOA) with five board members.*

Sample space

(a) *In how many ways can the HOA be formed?*

(b) *What is the probability that the HOA consists of 2 dogs and 3 cats?*

(c) *What is the probability that the HOA consists of at least 2 cats?*

Solution.

(a) *There are 35 residents, 5 will be chosen. The total number of ways this can be done is*

$$C(35,5) = 324632$$

(b) *There are $C(15,2)$ ways to have two dogs, $C(20,3)$ ways to have three cats. The total number of ways to have two dogs and three cats is, according to the multiplication principle, $C(15,2)C(20,3)$. The probability of having two dogs and three cats, by (5.2.1), is*

$$\Pr(2 \text{ dogs } 3 \text{ cats}) = \frac{C(15,2)C(20,3)}{C(35.5)} = \frac{119700}{324632} \approx 0.369$$

(c) *The event {at least 2 cats} includes four outcomes: "3 dogs 2 cats", "2 dogs 3 cats", "1 dog 4 cats" and "0 dogs 5 cats".*

$$\begin{aligned} \Pr(\text{at least 2 cats}) &= \Pr(3D2C) + \Pr(2D3C) + \Pr(1D4C) + \Pr(0D5C) \\ &= 1 - [\Pr(5D0C) + \Pr(4D1C)] \\ &= 1 - \left[\frac{C(15,5)C(20,0)}{(35,5)} + \frac{C(15,4)C(20,1)}{C(35,5)} \right] \approx 0.907 \end{aligned}$$

Here we used formula (5.2.2) again. Make sure you see which event is E and which event is E' in the formula. (5.2.2) is useful when E is substantially larger than E' ◀

Example 5.2.3. *Fifteen dogs, twenty cats, and ten opossums live in a non-gated community. They want to randomly form a patrol team of seven members.*

(a) *How many different patrol teams can be formed?*

(b) *What is the probability the patrol team consists of 3 dogs, 2 cats, and 2 opossums?*

(c) *What is the probability the patrol team consists of at least one opossum?*

Solution.

(a) *There are 45 residents, 7 will be chosen. The total number of ways this can be done is*

$$C(45, 7) = 45379620$$

(b) *Choose 3 dogs from 15, 2 cats from 20, and 2 opossums from 10. The total number of ways this can be done is $C(15, 3)C(20, 2)C(10, 2) = 3890250$.*

$$\Pr(3 \text{ dogs 2 cats 2 opossums}) = \frac{3890250}{45379620} \approx 0.086$$

(c) *Since the question concerns only opossums, we divide the residents into two groups: the first group consists of 10 opossums, and the second group consists of 35 non-opossums. The complement of {at least one opossum} is {no opossums} and the number of ways the team consists of no opossums is $C(10, 0)C(35, 7) = 6724520$.*

$$\Pr(\text{at least one opossum}) = 1 - \Pr(\text{no opossums})$$
$$= 1 - \frac{6724520}{45379620} \approx 0.852 \blacktriangleleft$$

This last example fully displays the beauty of formula (5.2.2), and the technique of combining several groups into one when we do not need to differentiate them. If we do not have these two tools at our disposal, here is what we have to do: Let (x, y, z) represent the number of dogs, number of cats, and number opossums in the team. To have at least one opossum, (x, y, z) can be

$$(6, 0, 1), (5, 1, 1), (4, 2, 1), (3, 3, 1), (2, 4, 1), (1, 5, 1), (0, 6, 1)$$
$$(5, 0, 2), (4, 1, 2), (3, 2, 2), (2, 3, 2), (1, 4, 2), (0, 5, 2)$$
$$(4, 0, 3), (3, 1, 3), (2, 2, 3), (1, 3, 3), (0, 4, 3)$$
$$(3, 0, 4), (2, 1, 4), (1, 2, 4), (0, 3, 4)$$
$$(2, 0, 5), (1, 1, 5), (0, 2, 5)$$
$$(1, 0, 6), (0, 1, 6)$$
$$(0, 0, 7)$$

There are 28 compositions, and we need to calculate each one and then add them together. That would be rather tiring. Sure we can do it if we have to, but we'd rather not if we don't have to.

Example 5.2.4. *A quiz consists of 10 True-or-False questions. Suppose a donkey randomly chooses an answer for each question (the word "randomly" implies the donkey is right with probability 0.5 and wrong also with probability 0.5).*

   (a) In how many different ways can the quiz be answered?
   (b) What is the probability that the donkey answers exactly two questions correctly?
   (c) What is the probability that the donkey answers five or more questions correctly?
   (d) What is the probability that the donkey answers at least one question correctly?

Solution.

   (a) *Each question has two possible answers, and there are 10 questions. The total number of ways to answer the quiz is*

$$2^{10} = 1024$$

   (b) *There are 10 questions, the two correct answers can be any two of the 10. The number of ways this can occur is $C(10,2) = 45$. The probability is*

$$\Pr(2 \ correct) = \frac{45}{1024} \approx 0.044$$

   (c) *Answering five or more questions correctly means answering exactly five, or six, or seven, or eight, or nine, or ten questions correctly*

$$\Pr(5 \ or \ more \ correct) = \Pr(5 \ c) + \Pr(6 \ c) + \Pr(7 \ c) + \Pr(8 \ c) + \Pr(9 \ c) + \Pr(10 \ c)$$
$$= \frac{C(10,5) + C(10,6) + C(10,7) + C(10,8) + C(10,9) + C(10,10)}{1024}$$
$$= \frac{638}{1024} \approx 0.623$$

I.e., the donkey can actually answer 50% or more questions correctly 62% of the time.

   (d) *Answering at least one question correctly is the complement of answering zero questions correctly*

$$\Pr(at \ least \ one \ correct) = 1 - \Pr(0 \ correct)$$
$$= 1 - \frac{C(10,0)}{1024} = \frac{1023}{1024} \approx 0.999$$

Amazing! It is almost certain the donkey will answer at least one question correctly◀

Before we go on, please make sure you see that the last example is another coin flipping problem in disguise.

:: **EXERCISES 5.2** ::

1. There are five red and six yellow marbles in a bowl. Find the probability that a randomly chosen marble is red.

2. Five boys and six girls play outside their homes. They accidentally break Mr. Robinson's porch light. One of them will be chosen randomly to tell Mr. Robinson what happened. Find the probability that a boy is chosen.

3. Three freshmen, four sophomores, and five seniors organized a fundraising event. After the event was over one person was randomly selected to take down all the banners. What is the probability that a freshman was selected?

4. Five freshmen and four sophomores start a book club. A committee of three will be randomly formed to brainstorm strategies on recruiting new members. Find the probability the committee consists of one freshman and two sophomores.

5. Billy the Bully gets hit from behind by three paper balls thrown by Zach, Joe, and Pete. He turns around and sees nine kids laughing. If Billy randomly picks three kids, what is the probability that he picks

    (a) The three guilty ones?

    (b) Two of the three guilty ones and an innocent one?

    (c) One of the three guilty ones and two innocent ones?

    (d) Three innocent ones?

    (e) At least one guilty one?

    (f) Verify and explain why the sum of the first four numbers above equals 1.

    (g) Verify and explain why the fifth number above equals one minus the fourth number.

    (h) Some of the fractions above could have been reduced. What is the benefit of not reducing them in this particular case?

6. Given six squares □□□□□□. Suppose we are to color each square either purple or blue, randomly.

    (a) How many different coloring sequences are possible?

    (b) What is the probability that two of the squares are colored purple and four are colored blue?

    (c) What is the probability that at least one of the squares is colored purple?

7. Flip a fair coin six times and record the sequence of heads and tails.

    (a) How many different sequences are possible?

    (b) Find the probability that a sequence consists of two heads and four tails.

    (c) Find the probability that a sequence consists of at least one heads.

    (d) If you feel déjà vu, look at the previous problem again.

8. Mrs. Sun's Pizza Grills offers six toppings on a 3-D triangular shaped pie. A Martian walks into Mrs. Sun's place and orders a pizza by going through the toppings one by one and randomly deciding whether a topping should be ordered or not.

    (a) In how many different ways can the Martian order a pizza?

    (b) What is the probability that the Martian orders a two-topping pizza?

    (c) What is the probability that the Martian orders a pizza that has at least one topping?

    (d) If you feel déjà vu all over again one more time all over again, make sure you see that the last two problems and this current one are mathematically equivalent. So here we have covered applications in three different areas: painting, gaming, and extraterrestrial contact.

9. Mary's beloved chicken lays one egg per day. Each egg can be white, yellow, or blue, each color occurring with probability 1/3. What is the probability that the chicken laid one white, two yellow, and four blue eggs last week?

10. A long time ago on an island far, far away lived 47 pirates. 13 of them were vegans, 17 of them were rum drinkers, and 5 of them were vegans and rum drinkers. Suppose a pirate was randomly chosen. What is the probability that this pirate

    (a) Was a rum drinker?

    (b) Was a vegan?

    (c) Was a vegan and a rum drinker?

    (d) Was either a vegan or a rum drinker?

    (e) Was neither a vegan nor a rum drinker?

11. HOAs (Home Owners Associations) can really make home owners anxious with their Notice of Violations letters. 53% of home owners who received such letters experienced nervousness, 67% of home owners who received such letters experienced outrage, and only 1% experienced neither of the two aforementioned effects. What is the percentage of home owners who experienced both effects?

5.3 Conditional Probability and Independent Events

In general, the sample space of an experiment does not have to be finite. For example, if we flip a coin and record the number of flips before the first heads appears, then the sample space is $\{0, 1, 2, 3, 4, ...\}$, i.e., all the nonnegative integers, because we just don't know when the first heads will appear.

In this section we will present some principles and formulas that apply to more general situations beyond the ones that can be solved by the counting methods as presented in the last section.

First we introduce the notion of *conditional probability*. Suppose E and F are two events in the same sample space. The conditional probability of E given F is defined as

---

Conditional probability of E given F

$$\Pr(E|F) = \frac{\Pr(E \cap F)}{\Pr(F)} \qquad (5.3.1)$$

Let's look at an example first. Suppose in a parking lot there are 18 cars, 10 of them are Chevrolets and 8 of them are Fords, 6 of the Chevrolets and 5 of the Fords are red; the other cars are black. See the Venn diagram below.

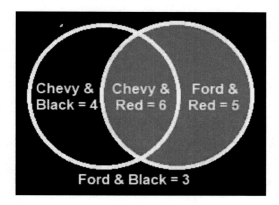

If we randomly pick a car, the probability of picking a Chevrolet is obviously 10/18. If we randomly pick a car and see that it is red, and after that we want to know the probability that this car is a Chevrolet, the probability is changed to 6/11 because there are 11 red cars and 6 of them are Chevrolets. 10/18 is the probability of picking a Chevrolet, 6/11 is the conditional probability of picking a Chevrolet given that the car picked is red.

$$\Pr(\text{Chevy}) = \frac{10}{18}$$

$$\Pr(\text{Chevy}|\text{Red}) = \frac{6}{11}$$

From the Venn diagram we see that the numerator 6 in $\Pr(\text{Chevy}|\text{Red}) = \frac{6}{11}$ is the number of red Chevys, and the denominator 11 is the number of red cars in the sample space. Let's make sure definition (5.3.1) will give us the same answer:

$$\Pr(\text{Chevy}|\text{Red}) = \frac{\Pr(\text{Chevy \& Red})}{\Pr(\text{Red})} = \frac{6/18}{11/18} = \frac{6}{11}$$

This example provides a concrete justification for the definition (5.3.1). The formula is quite easy to memorize once you see the reasoning behind it: The given condition F becomes the new sample space because we know it has occurred, that's why $\Pr(F)$ is in the denominator, and since F is now the sample space, only the part of E that is in F, i.e., $E \cap F$, matters, and that's why $\Pr(E \cap F)$ is in the numerator.

Example 5.3.1. *Toss a fair coin 5 times and record the sequence of heads and tails.*

  (a) Find the total number of outcomes.
  (b) Find the number of sequences that contain exactly two heads and three tails.
  (c) Find the probability that a sequence contains two heads and three tails.
  (d) If the first toss is heads, find the probability that the sequence contains exactly two heads.

Solution.

  (a) Every toss can result in heads or tails, that's 2 outcomes, repeated 5 times. The total number of outcomes is $2^5 = 32$.
  (b) Five spots, pick two for heads. The answer is $C(5,2) = 10$.
  (c) Use the numbers obtained above

$$\Pr(2H\,3T) = \frac{10}{32} = \frac{5}{16}$$

  (d) Let F be the event {the first toss is heads}. It is obvious that $\Pr(F) = 1/2$ since the coin is fair. Let E be the event {there are two heads in the sequence}. We are looking for the conditional probability $\Pr(E|F)$. Follow formula (5.3.1), we know we need to find $\Pr(E \cap F)$.

  $E \cap F = \{$the first toss is H, and there is one additional H in the remaining four tosses$\}$. $E \cap F$ is one event that involves two steps: tossing heads first followed by tossing heads one more time in the next four tosses. The number of ways this can occur is $C(1,1)C(4,1) = 4$. So

$\Pr(E \cap F) = 4/32 = 1/8$. *And*

$$\Pr(E|F) = \frac{\Pr(E \cap F)}{\Pr(F)} = \frac{1/8}{1/2} = \frac{1}{4} \blacktriangleleft$$

Example 5.3.2. *75% of freshmen at a college have taken English 101, 50% have taken math 101, and 35% have taken both. A freshman is randomly chosen.*

(a) *What is the probability that the student has taken at least one of the two courses?*
(b) *What is the probability that the student has taken neither course?*
(c) *Given that the student has taken English 101, what is the probability that the student has also taken math 101?*
(d) *Given that the student has taken math 101, what is the probability that the student has also taken English 101?*

Solution. *Let E represent the set of freshmen who have taken English 101, and M the set of freshmen who have taken math 101. We are given* $\Pr(E) = 0.75$, $\Pr(M) = 0.50$, *and* $\Pr(E \cap M) = 0.35$.

(a) *"Taken at least one of the two courses" is the set* $E \cup M$. *Using (5.2.3) we find*

$$\Pr(E \cup M) = \Pr(E) + \Pr(M) - \Pr(E \cap M) = 0.75 + 0.50 - 0.35 = 0.90$$

The probability is 0.90 that a freshman has taken at least one of the two courses.

(b) *"Taken neither course" is the complement of "taken at least one of the two courses". Using (5.2.2) we find*

$$\Pr(\text{taken neither course}) = 1 - \Pr(E \cup M) = 1 - 0.90 = 0.10$$

(c) *We know the student has taken English 101, so E is the given condition, and we are looking for the probability that under this given condition, the student has also taken math 101, i.e.,* $\Pr(M|E)$. *Using (5.3.1) we find*

$$\Pr(M|E) = \frac{\Pr(E \cap M)}{\Pr(E)} = \frac{0.35}{0.75} = \frac{7}{15} \approx 0.467$$

(d) *This one is easy after the last one. We are looking for*

$$\Pr(E|M) = \frac{\Pr(E \cap M)}{\Pr(M)} = \frac{0.35}{0.50} = \frac{7}{10} = 0.70 \blacktriangleleft$$

If we change $\Pr(E \cap M)$ from 0.35 to 0.375 in the example above, something interesting happens. We notice that

$$\Pr(M|E) = \frac{0.375}{0.75} = 0.50 = \Pr(M)$$

and

$$\Pr(E|M) = \frac{0.375}{0.50} = 0.75 = \Pr(E)$$

i.e., the given condition of an event does not change the probability of the event. When this happens, we say that the two events are *independent*.

> Independent events. Two events E and F are independent if and only if
>
> $$\Pr(E|F) = \Pr(E) \quad \text{or} \quad \Pr(F|E) = \Pr(F) \qquad (5.3.2)$$

From (5.3.1) we see that $\Pr(E \cap F) = \Pr(E|F)\Pr(F)$. If E and F are independent, then $\Pr(E \cap F) = \Pr(E)\Pr(F)$. This equation can also be used as the definition of two events being independent.

> Independent events. Two events E and F are independent if and only if
>
> $$\Pr(E \cap F) = \Pr(E)\Pr(F) \qquad (5.3.3)$$

Either (5.3.2) or (5.3.3) can be used to verify whether two events are independent. Remember this is a mathematical definition; it is not subject to arguments. If the probabilities of two events and their intersection are known or can be derived from known information, we must use the definition to check whether these two events are independent. One cannot argue loudly and say "I believe so" or in general bully other people into agreeing that the two events are independent or not independent.

Example 5.3.3. *Every morning with probability* 0.60 *a math teacher living on a mountain top by himself, without any connection to the outside world, will eat a hare. Every morning with probability* 0.70 *the president will drink two cups of coffee. The probability that these two things happen in the same morning is* 0.40. *Are these two events independent?*

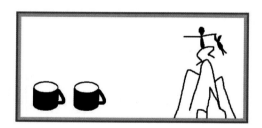

The Strong, but Wrong, Solution. *Whatever this lonely math teacher character does every morning has nothing to do with what the president does. It is so obvious only some superstitious person would believe these two things are not independent. If you think these two things are not independent, you've got some serious logical reasoning issues with your brain. You are illogical and you need to get logicalized.*

The Light, but Right, Solution. *Let E be the event "the math teacher eats a hare" and F the event "the president drinks two cups of coffee". We know* $\Pr(E) = 0.60$, $\Pr(F) = 0.70$, *and* $\Pr(E \cap F) = 0.40$. *Since* $\Pr(E)\Pr(F) = 0.60 \times 0.70 = 0.42 \neq \Pr(E \cap F)$, *according to* (5.3.3), E *and* F *are NOT independent*◄

The moral of the story: The word "independent", when used in the probability context, has a strict mathematical definition (just like the word "or" when used in sets indicates the union operation). When a common word is redefined in a mathematical definition, we must not allow the "everyday" interpretation of the word to interfere with our judgment. The word has been redefined and we need to have the discipline to keep the new definition in mind. For example, you have an understanding what the word "fish" means. But if you go to another country and in that country "fish" stands for all animals that can move themselves using fins or wings, then you need to keep that in mind when you are in that country and know that the owl is now a member of the "fish" family.

It turns out that generalizing (5.3.3) to a formula that applies to three or more events is not that trivial. We will simply point out here that *if three or more events are mutually independent, then the probability of their intersection equals the product of the probabilities of the individual events.* Be careful though that the converse of this statement is not true, i.e., if the product of the probabilities of three or more events equals the probability of the intersection of these events, it is not a guarantee that these events are mutually independent.

Example 5.3.4. *One module in a machine consists of three identical, mutually independent components. The module fails only if all three components fail. Suppose the probability that a component fails in 2000 hours of operation is 0.01.*

(a) *Find the probability that the module fails in 2000 hours of operation.*
(b) *Find the probability that the module does not fail in 2000 hours of operation.*

Solution. *Let F_1, F_2, and F_3 represent the events "component 1 fails in 2000 hours of operation", "component 2 fails in 2000 hours of operation", and "component 3 fails in 2000 hours of operation", respectively. We are given $\Pr(F_1) = \Pr(F_2) = \Pr(F_3) = 0.01$.*

(a) *Let F be the event "the module fails". F occurs when "component 1 fails, component 2 fails, and component 3 fails", i.e., $F = F_1 \cap F_2 \cap F_3$. Remember that F_1, F_2, and F_3 are mutually independent, it follows that*

$$\Pr(F) = \Pr(F_1)\Pr(F_2)\Pr(F_3) = 0.01 \times 0.01 \times 0.01 = 0.000001$$

(b) *The event "the module does not fail" is F', the complement of F,*

$$\Pr(F') = 1 - \Pr(F) = 0.999999 \blacktriangleleft$$

Example 5.3.5. *You told three friends A, B, and C, independently, what you did last summer and they all promised to keep the secret for you. However, with probability 0.3 A will reveal your secret to the world, with probability 0.2 B will reveal your secret to the world, and with probability 0.1 C will do the same. What is the probability that the world will know what you did last summer?*

Solution. *The world will know your secret if at least one the three friends reveals your secret. The* *complement of "at least one reveals" is "no one reveals". Notice that the probabilities that A, B, and C don't reveal the secret are* 0.7, 0.8, *and* 0.9, *respectively. So*

$$\Pr(\textit{the world knows}) = 1 - \Pr(\textit{no one reveals}) = 1 - 0.7 \times 0.8 \times 0.9 = 0.496$$

That's almost 50%! *We sincerely hope you didn't do something really bad last summer*◄

:: **EXERCISES 5.3** ::

1. The conditional probability of E given F is denoted by the notation _____ .

2. $\Pr(E|F) =$

     (A) $\dfrac{\Pr(E)}{\Pr(F)}$ (B) $\dfrac{\Pr(E \cap F)}{\Pr(F)}$ (C) $\dfrac{\Pr(E \cap F)}{\Pr(E)}$ (D) $\Pr\left(\dfrac{E}{F}\right)$

3. Let D be the event "a randomly chosen voter is a Democrat", R be the event "a randomly chosen voter is a Republican", and P be the event "a randomly chosen voter voted for Paul". Use the appropriate probability notations to express the following probabilities (for example, the probability that a randomly chosen voter is a Democrat who voted for Paul is denoted by $\Pr(D \cap P)$):

     (a) The probability that a randomly chosen voter is a Democrat given that this voter voted for Paul.

     (b) The probability that a randomly chosen voter voted for Paul given that this voter is a Republican.

     (c) The probability that a randomly chosen voter is a Republican and voted for Paul.

     (d) The probability that a randomly chosen voter is a Republican and a Democrat.

     (e) The probability that a randomly chosen voter is a Republican or a Democrat.

     (f) The probability that a randomly chosen voter is a Democrat given that voter is a Republican.

4. Suppose E and F are two events, with $\Pr(E) = 0.6$, $\Pr(F) = 0.4$, $\Pr(E \cap F) = 0.2$. Find

     (a) $\Pr(E|F)$ (b) $\Pr(F|E)$

5. Suppose E and F are two events, with $\Pr(E) = 0.7$, $\Pr(F) = 0.5$, $\Pr(E \cup F) = 0.9$. Find

    (a) $\Pr(E \cap F)$

    (b) $\Pr(E|F)$

    (c) $\Pr(F|E)$

    (d) $\Pr(F'|E)$ Hint: $\Pr(F' \cap E) + \Pr(F \cap E) = \Pr(E)$

    (e) $\Pr(F|E')$

    (f) $\Pr(F'|E')$

    (g) Is it always true that $\Pr(F|E) = 1 - \Pr(F'|E)$?

    (h) Are E and F independent? Justify your answer.

6. Hui-An is a small town sitting by the Pacific Ocean. 40% of the families in the town are in the fishing business, 30% are in the construction business, and 12% of families are in both. A family is randomly chosen.

    (a) What is the probability that this family is in neither the fishing nor the construction business?

    (b) What is the probability that this family is in the construction business if it is known that the family is in the fishing business?

    (c) What is the probability that this family is in the fishing business if it is known that the family is in the construction business?

    (d) Are the events "in the fishing business" and "in the construction business" independent? Justify your answer.

7. In the self-claimed Happiest City in the World, 40% of the citizens drink, 35% of the citizens gamble, and 55% of the citizens neither drink nor gamble. A citizen is randomly chosen.

    (a) What is the probability that this citizen drinks and gambles?

    (b) What is the probability that this citizen gambles given that this citizen drinks?

    (c) What is the probability that this citizen drinks given that this citizen gambles?

    (d) Are the events "drinking" and "gambling" independent? Justify your answer.

8. Dr. Stubborn is a research scientist who specializes in the field of quantum conspiracy theory. With full confidence in his own scientific intuition, he has always argued and threatened everyone into agreeing with him that the event "his round shaped fish breakdances in its tank during the night" and the event "Sherman the Monkey sneezes in his sleep" are independent. Long story short, Dr. Stubborn's Ph.D. student Dr. Sleepin couldn't find a job after graduation and moved in with him. Dr. Sleepin slept in the living room couch and had the chance to observe the round shaped fish and Sherman the Monkey during his numerous sleepless nights. After observing these two wonderful creatures for over ten years, Dr. Sleepin has concrete evidence that with probability 0.3 the round shaped fish

breakdances on any given night, with probability 0.2 Sherman the Monkey sneezes in his sleep on any given night, and the two events occur on the same night with probability 0.15. In light of this new evidence, should Dr. Stubborn change his mind about the two events being independent if he is truly a man of science as he has always claimed?

9. A robot consists of 50 components, and all the components function independently. Suppose the probability that any single component fails during the first year of the robot's operation is 0.0001.

   (a) What is the probability that the robot will live its first year trouble-free? Round answer to the nearest thousandths place.

   (b) Suppose 10 more components are added to the robot and all conditions remain the same. What is the probability that the robot will live its first year trouble-free?

   (c) Suppose the robot is redesigned and now consists of only 30 components, and assume all other conditions remain the same. What is the probability that the robot will live its first year trouble-free?

   (d) Based on the calculations above, we can make a conjecture that the more components a system consists of, the (circle one) MORE / LESS trouble the system is going to have.

   (e) Reflect on the last question-answer.

10. Three chimpanzees are given the same puzzle to solve. Each chimpanzee is situated in a separate room so they can't see their peers' solutions. Suppose chimpanzee 1 can solve the puzzle with probability 0.9, chimpanzee 2 can solve the puzzle with probability 0.8, and chimpanzee 3 can solve the puzzle with probability 0.7.

   (a) What is the probability that no chimpanzee solves the puzzle?

   (b) What is the probability that exactly one chimpanzee solves the puzzle?

   (c) What is the probability that exactly two chimpanzees solve the puzzle?

   (d) What is the probability that all three chimpanzees solve the puzzle?

   (e) What is the probability that at least one chimpanzee solves the puzzle?

5.4   TREE DIAGRAMS AND BAYES' THEOREM

A tree diagram is another way to solve some probability problems. A tree diagram works best when the total number of outcomes is not too large, and the experiment does not involve too many steps.

Example 5.4.1. *There are* 8 *red and* 5 *green marbles in a bag. We randomly draw a marble from the bag. If the marble is red, the experiment stops; if the marble is green, we keep the marble and draw a second marble. If the second marble is red, the experiment stops; if it is green, we keep the second marble and draw a third marble. The experiment stops after the third marble is drawn regardless of its color.*

  (a) *What is the probability that the first marble is red?*
  (b) *What is the probability that the experiment ends after the second marble is drawn?*
  (c) *What is the probability that the experiment ends after the third marble is drawn?*
  (d) *What is the probability that we draw exactly two green marbles in the experiment?*

Solution. *A tree diagram is suitable in this case because the experiment does not fit into any of the models we have studied so far. To draw a tree diagram, start from one point and draw branches that lead to all the possible outcomes in the first step, one branch for each outcome, and write the corresponding probability on the branch. If a first-step outcome has more outcomes that follow, we then draw branches from this outcome to all the possible outcomes that follow, and put the corresponding probabilities on the branches. Continue this process until all possible outcomes have been accounted for.*

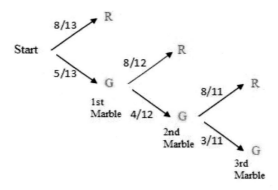

Let's look at the tree and the numbers. From the Start position, there are 8 *red and* 5 *green marbles in the bag, for a total of* 13 *marbles. So when we draw one marble, the probability of drawing a red is* 8/13 *and a green* 5/13. *If the first marble is red the experiment stops, that's why there are no more branches following the first R; if the first marble is green, a second marble is drawn but since the first marble is not returned to the bag, there are now* 8 *red and* 4 *green marbles in the bag, and the probabilities change accordingly,* 8/12 *for drawing red and* 4/12 *for drawing green. The reader should make sense of the numbers on the last two branches.*

  (a) Pr(*the first marble is red*) = 8/13.
  (b) *The experiment ends after the second marble is drawn if the first marble is green and the second marble is red. So from Start we go to G, and then from that G we go to R. The probabilities on these two branches are* 5/13 *and* 8/12, *respectively. We multiply them together to get the final probability* (5/13)(8/12) = 10/39. *To see why we multiply the two*

probabilities together, let's we go back to (5.3.1).

$$\Pr(first\ G\ AND\ second\ R) = \Pr(first\ G)\,\Pr(second\ R|first\ G) = (5/13)(8/12) = 10/39$$

(c) *The experiment ends after the third marble is drawn if the first marble is green and the second marble is also green. So we go from Start to G to G, the probability of this happening is (5/13)(4/12) = 5/39.*

(d) *The only way that we draw exactly two green marbles in the experiment is if the first two marbles are green and the third one is red. So we go from Start to G to G to R, the probability is (5/13)(4/12)(8/11) = 40/429* ◄

When drawing a tree, it is important to concentrate on outcomes rather than physical objects. In our last example, it would have been a mistake had we drawn 13 branches from the Start position, with 8 branches each leading to a red marble and 5 branches each leading to a green marble. Try that and you will see how the tree grows big really fast and the answers we are looking for become difficult to find.

Example 5.4.2. *Suppose there are five freshmen, four sophomores, and three juniors in a room. Two freshmen, three sophomores, and one junior are from the state of Tennessee. One student is randomly chosen.*

(a) *What is the probability that this student is a freshman?*

(b) *What is the probability that this student is a freshman from Tennessee?*

(c) *What is the probability that this student is from Tennessee?*

(d) *Given that the student is from Tennessee, what is the probability that the student is a freshman?*

Solution. *We can try to list all the students but it will be inefficient, because we will have to say something like "the five freshmen are Freshman 1, Freshman 2, Freshman 3, Freshman 4 and Freshman 5, and Freshman 1 and Freshman 2 are from Tennessee." Instead we will use a tree diagram to analyze the numbers.*

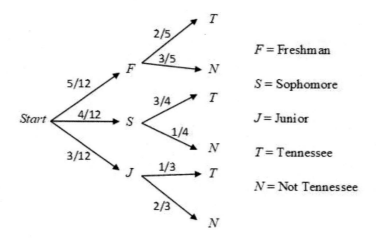

(a) $\Pr(F) = 5/12$ *because there are 12 students and 5 of them are freshmen. This is the number on the branch from Start to F.*

(b) $\Pr(F\ AND\ T) = (5/12)(2/5) = 1/6$. *We go from Start to F first to get a freshman, then from F to T to get a freshman who is from Tennessee.*

(c) *Here we have to find all sequences that end with a T. The sequences are Start-F-T, Start-S-T, and Start-J-T. These sequences are mutually exclusive, so*

$$\Pr(T) = \Pr(F \cap T) + \Pr(S \cap T) + \Pr(J \cap T)$$
$$= (5/12)(2/5) + (4/12)(3/4) + (3/12)(1/3) = 1/2$$

(d) *We are looking for*

$$\Pr(F|T) = \frac{\Pr(F \cap T)}{\Pr(T)} = \frac{(1/6)}{(1/2)} = \frac{1}{3} \blacktriangleleft$$

The conditional probability in the last example belongs to a category of problems called posterior probability problems. Let's revisit the example and investigate all the characteristics in the setup of the problem.

1. The sample space is divided into several mutually exclusive subsets. In this example, the sample space is all the students in the room, and the mutually exclusive subsets are {*Freshmen*}, {*Sophomores*}, and {*Juniors*}. We say these subsets *partition the sample space.*

2. There is a subset that intersects with all the partitioning subsets. In this example, this is the subset {*Tennessee*}. Each of the three subsets contains some elements in {*Tennessee*}. We call this subset the *set of interest.*

3. The probabilities of all the subsets that partition the sample space are known. In this example, we know $\Pr(F) = 5/12$, $\Pr(S) = 4/12$, and $\Pr(J) = 3/12$.

4. The conditional probability of the set of interest given any partitioning subset is known. In this example we know $\Pr(T|F) = 2/5$, $\Pr(T|S) = 3/4$, and $\Pr(T|J) = 1/3$.

5. We are interested in the probability of a partitioning subset given the set of interest. In this

example we want to find the probability of choosing a freshman given that it is already known the student is from Tennessee, i.e., $\Pr(F|T)$.

Using the set operations $\cap$ and $\cup$ for the connective words "and" and "or", respectively, we can express the conditional probability $\Pr(F|T)$ in terms of the given probabilities:

$$
\begin{aligned}
\Pr(F|T) = \frac{\Pr(F \cap T)}{\Pr(T)} &= \frac{\Pr(F \cap T)}{\Pr((F \cap T) \cup (S \cap T) \cup (J \cap T))} \\
&= \frac{\Pr(F \cap T)}{\Pr(F \cap T) + \Pr(S \cap T) + \Pr(J \cap T)} \\
&= \frac{\Pr(F)\Pr(T|F)}{\Pr(F)\Pr(T|F) + \Pr(S)\Pr(T|S) + \Pr(J)\Pr(T|J)}
\end{aligned}
$$

The last expression gives us the Bayes' theorem or Bayes' formula. The following is Bayes' theorem when the sample space is partitioned into two mutually exclusive events F_1 and F_2.

> **Bayes' Theorem**
> $$\Pr(F_1|E) = \frac{\Pr(F_1)\Pr(E|F_1)}{\Pr(F_1)\Pr(E|F_1) + \Pr(F_2)\Pr(E|F_2)} \qquad (5.4.1)$$

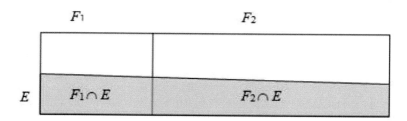

The Bayes' formula has this intimidating look, which is good. If you are really into math, then you should get yourself totally comfortable with the formula's presentation, and be able to see the meaning behind the formula. If you are not that into math, then remember you can still solve most of the Bayes' problems in this book if you have a good understanding of the example we have discussed. Just draw a diagram like the one shown above and the solution will be easy to find.

Example 5.4.3. *Your friend just won the Powerball jackpot and is throwing an orange party (what a cheapskate). You are the first one to arrive. You learned that 80% of the oranges at the party are from California, and 20% are from Arizona. Of the California oranges, 90% are sweet, while 95% of the Arizona oranges are sweet. You randomly pick an orange and eat it, and it is sweet. What is the probability that the orange is from Arizona?*

Solution I. *We will first use the more visual approach. Let's draw a partition diagram and fill the regions in the diagram with numbers.*

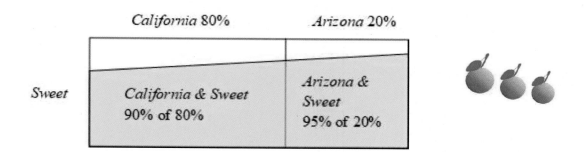

The question asks for Pr(*Arizona*|*Sweet*). *The given condition is Sweet, which is the shaded area in the lower part of the diagram.*

$$\text{Pr}(Arizona|Sweet) = \frac{\text{Pr}(Arizona \ \& \ Sweet)}{\text{Pr}(Sweet)} = \frac{95\% \cdot 20\%}{90\% \cdot 80\% + 95\% \cdot 20\%} \approx 0.21$$

Solution II. *We can use formula* (5.4.1) *to find the answer without relying on a diagram. Use C, A, and S to denote the events* {*California*}, {*Arizona*}, *and* {*Sweet*}, *respectively. We are given* $\text{Pr}(C) = 0.80$, $\text{Pr}(A) = 0.20$, $\text{Pr}(S|C) = 0.90$, *and* $\text{Pr}(S|A) = 0.95$. *We are looking for*

$$\text{Pr}(A|S) = \frac{\text{Pr}(A)\,\text{Pr}(S|A)}{\text{Pr}(A)\,\text{Pr}(S|A) + \text{Pr}(C)\,\text{Pr}(S|C)} = \frac{0.20 \times 0.95}{0.20 \times 0.95 + 0.80 \times 0.90} = \frac{19}{91} \approx 0.21 \blacktriangleleft$$

The example above can also be solved with a tree diagram, similar to the Students/Tennessee example (find the example now, if you don't remember it).

Example 5.4.4. *On this island somewhere in the Pacific Ocean, 30% of the people are convicted of committing crime Z. Of those convicted, 10% are innocent. Of the 70% not convicted, 5% are guilty. If a man on this island confesses to a priest that he is guilty of committing crime Z, what is the probability that this man has not been convicted of committing crime Z?*

Solution. *Well, no full solution actually. The answer is* 7/61, *or about* 0.11. *An incomplete diagram is provided below. You need to fill the regions with numbers, and then verify the given answer. Please do.* Justice is in your hands!

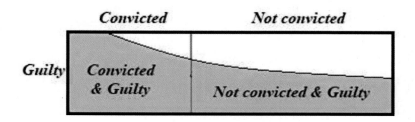

:: **EXERCISES 5.4** ::

1. For the next vacation the Robinsons are considering three destinations: the Grand Canyon, Catalina Island, and Las Vegas. The probabilities that they choose these destinations are 0.5, 0.3, and 0.2, respectively. Draw a tree diagram that shows all possible vacation destinations and their corresponding probabilities.

2. You are stuck in a traffic jam on a highway in this great golden state. You would like to eat in fifteen minutes. With probability 0.75 the traffic won't improve and you can find an exit in fifteen minutes where you can have a hotdog or a piece of microwaved pizza, each with probability 0.5. With probability 0.25 the traffic will ease up and you can get to the next town in fifteen minutes where you know you can find a Mexican restaurant, a spicy Thai restaurant, and a good old American restaurant, and the probabilities that you choose these restaurants are 0.65, 0.20, and 0.15, respectively.

    (a) Draw a tree diagram that displays all the possible options of your next meal, and their probabilities.

    (b) What is the probability that you will have a hotdog as your next meal?

    (c) What is the probability that you will have spicy Thai food as your next meal?

3. It is December and you have applied to three universities (the same three your older sister applied to just because you wanted to show people that you can, too) and are waiting for their replies. University A was your first choice, meaning that if A accepts you then you will definitely go to A. University B is your second choice so if A rejects you but B accepts you then you will definitely go to B. University C is your third choice so you will go to C if A and B both reject you and C accepts you. If all three colleges reject you, you will enroll at the College of Southern Nevada (CSN). The probabilities that universities A, B and C will accept you are 0.2, 0.3, and 0.4, respectively.

    (a) Draw a tree diagram that lays out all the education paths you could be taking next year, and their probabilities.

    (b) What is the probability that you will go to University A?

    (c) What is the probability that you will go to University B? (It is not 0.3)

    (d) What is the probability that you will be enrolled at CSN? (It is not 0.1)

    (e) What is the probability that you will go to one of the three universities?

4. After a spring break party Mr. Governor comes home at 4 A.M. He has seven keys in his pocket and even though the keys are all different Mr. Governor is unable to tell the differences at this point. One of the keys opens the front door. Mr. Governor decides to try the keys by the trial-and-error method he learned at school. If a key opens the front door, great! If a key doesn't open the front door, he throws it at his neighbor's window with all his strength to make sure the useless key doesn't return to his pocket.

    (a) Draw a tree diagram that illuminates all the possible key-trying sequences Mr. Governor may experience.

    (b) What is the probability that the first key Mr. Governor tries opens the front door (the neighbor will be very thankful)?

    (c) What is the probability that the second key opens the front door?

    (d) What is the probability that the seventh key opens the front door (the poor neighbor's window would have been broken six times)?

    (e) Find, without performing any actual calculation, the probability that the third, the fourth, the fifth, or the sixth key opens the front door.

5. Seven people are given seven identical envelopes. One of the envelopes contains three tickets to the Disneyland, the other six are empty. They will pick the envelopes based on seniority. The oldest person gets to pick an envelope first. If this envelope contains the tickets, the drawing is over and everyone goes home. If the first envelope is empty, the second oldest person then picks an envelope from the remaining six and checks if this envelope contains the tickets. This process continues until one person picks the envelope that has the tickets. The youngest person feels mistreated and wants to reverse the order. The oldest person gets upset that the youngest is not compliant so they get into a heated argument. Your job is to convince them, with solid mathematical evidence, that there is no difference either way they go.

6. 70% of the propane tanks used by a company are from factory A, the other 30% are from factory B. Of the tanks from A 4% are defective, and from B, 8%. If a randomly selected tank is found to be defective, what is the probability that the tank is from factory A?

7. A study by some students found that in this wild animal park, 40% of the kangaroos are male and 60% are female. 70% of the male kangaroos are left-handed while 30% of the females are left-handed. If a kangaroo is randomly captured and found to be left-handed, what is the probability that this kangaroo is a female?

8. The flu season is upon us. For those who receive the flu vaccine, 30% will still catch the flu. For those who don't receive the flu vaccine, 45% will catch the flu. Suppose 25% of the population received the flu vaccine.

    (a) If a randomly chosen person had the flu, what is the probability that this person had received the flu vaccine?

(b) If a randomly chosen person didn't have the flu, what is the probability that this person did not receive the flu vaccine?

9. Three big auto makers A, B, and C have 45%, 35%, and 20% of the market share, respectively, in this part of the world. A's products consist of 40% pickup trucks and 60% passenger cars, B's products consist of 50% pickup trucks and 50% passenger cars, and C's products consist of 70% pickup trucks and 30% passenger cars. If a randomly selected vehicle is a pickup truck, what is the probability that it is made by C?

10. At this moment, deep, deep in the Pacific Ocean, a huge pile of bottles is sitting there, emitting toxins. 65% of these bottles were produced by company X, 25% by company Y, and 10% by company Z. Of the bottles produced by X, 90% were plastic; Y, 80%; and Z, 70%. If a shark randomly eats a plastic bottle, what is the probability that the bottle was produced by company X? By company Y? By company Z?

11. There are two charity organizations in a city. Charity A spends 90% of all donations it receives on administrative spending while Charity B spends 30% of all donations it receives on administrative spending. Suppose the residents of the city gave 80% of their donations to A and 20% to B last year.

(a) What percent of the donations in the city was used towards administrative spending?

(b) If I received $100 of charity assistance last year, how much of it came from B?

(c) Convince yourself this problem is algebraically equivalent to the Bayes' formula.

Chapter 6

The Expected Value and Other Statistics Measures

Probability and statistics are a little bit of an outlier in mathematics. The Gauss-Jordan method for solving a system is quite fancy a method, but when we see a system of equations we need to solve, we know that's the go-to method. What makes probability and statistics challenging is we need to recognize the types of problems we are facing and choose the appropriate methods or formulas.

6.1 Probability Distributions

Suppose an experiment has a finite number of outcomes. We use X to denote the outcome of the experiment. X is called the *random variable* associated with the experiment. If the sample space of the experiment is $S = \{x_1, x_2, \ldots, x_k\}$, then X can be any one of the outcomes in the sample space. If with probability p_i the experiment results in outcome x_i, we write $\Pr(X = x_i) = p_i$, or simply $\Pr(x_i) = p_i$, $i = 1, 2, \ldots, k$. The list of all the outcomes and their corresponding probabilities is called the *probability distribution* of the random variable X.

Random Variable X	Probability
x_1	p_1
x_2	p_2
$\vdots$	$\vdots$
x_k	p_k

Example 6.1.1. *A fair coin is tossed three times and the sequence of heads and tails is recorded. Let X be the number of heads in the sequence. Find the probability distribution of X.*

Solution. *The number of heads can be* 0, 1, 2, *or* 3. *The sample space is* $S = \{0, 1, 2, 3\}$. *There are* 8 *sequences (because* $2^3 = 8$*) of heads (H) and tails (T).* $X = 0$ *if the sequence is* TTT, *which occurs with probability* $C(3, 0)/8 = 1/8$. $X = 1$ *if the sequence consists of one heads and two tails (such as* THT*), which occurs with probability* $C(3, 1)/8 = 3/8$. $X = 2$ *if the sequence consists of two heads and one tails (such as* THH*), which occurs with probability* $C(3, 2)/8 = 3/8$. *Finally* $X = 3$ *if the sequence is* HHH, *which occurs with probability* $C(3, 3)/8 = 1/8$. *The probability distribution of* X *is therefore*

X = number of heads	Probability
0	1/8
1	3/8
2	3/8
3	1/8

Notice the sum of all the probabilities is always equal to 1: $1/8 + 3/8 + 3/8 + 1/8 = 1$ ◀

Example 6.1.2. *(See example 5.2.2 in section 5.2) Fifteen dogs and twenty cats live in a gated community. They want to randomly form a HOA with five board members. Let X be the number of dogs on the HOA board. Find the probability distribution of X.*

Solution. X *can be* 0, 1, 2, 3, 4, *or* 5. *For instance,* $X = 1$ *means there is* 1 *dog and* 4 *cats on the board.*

$$\Pr(X = 0) = \frac{C(15, 0)C(20, 5)}{C(35, 5)} = \frac{15504}{324632}$$

$$\Pr(X = 1) = \frac{C(15, 1)C(20, 4)}{C(35, 5)} = \frac{72675}{324632}$$

$$\Pr(X = 2) = \frac{C(15, 2)C(20, 3)}{C(35, 5)} = \frac{119700}{324632}$$

$$\Pr(X = 3) = \frac{C(15, 3)C(20, 2)}{C(35, 5)} = \frac{86450}{324632}$$

$$\Pr(X = 4) = \frac{C(15, 4)C(20, 1)}{C(35, 5)} = \frac{27300}{324632}$$

$$\Pr(X = 5) = \frac{C(15, 5)C(20, 0)}{C(35, 5)} = \frac{3003}{324632}$$

After converting the fractions to decimal approximations, we obtain the following probability distribution of X

$X = $ number of dogs	Probability
0	0.048
1	0.224
2	0.369
3	0.266
4	0.084
5	0.009

The reader may want to verify that the sum of all the probabilities is indeed 1 ◀

Example 6.1.1 is an example of a *binomial distribution*: Suppose an experiment has two outcomes, one called *success* that occurs with probability p, the other called *failure* that occurs with probability $q = 1-p$. The experiment is repeated n times (each repetition is called a *Bernoulli trial*), independently. Let X be the number of successes in these n repetitions. We have the following formula

---

The Binomial Distribution with parameters n and p (where $q = 1-p$ in the formula)

$$\Pr(X = k) = C(n, k)p^k q^{n-k}, \qquad k = 0, 1, \ldots, n \qquad (6.1.1)$$

Example 6.1.3. *(Same as example 6.1.1, but solved with (6.1.1)) You toss a fair coin three times. Let X be the number of heads in the sequence. Find the probability distribution of X.*

Solution. *Let's call heads "success" and tails "failure". Since the coin is fair, the probability of success is $p = 1/2$, and the probability of failure is $q = 1-p = 1/2$. The coin is tossed three times so $n = 3$. From (6.1.1), we find the probability distribution of X to be:*

$$\Pr(X = 0) = C(3, 0)(1/2)^0 (1/2)^3 = 1/8$$
$$\Pr(X = 1) = C(3, 1)(1/2)^1 (1/2)^2 = 3/8$$
$$\Pr(X = 2) = C(3, 2)(1/2)^2 (1/2)^1 = 3/8$$
$$\Pr(X = 3) = C(3, 3)(1/2)^3 (1/2)^0 = 1/8$$

Which is of course the same as the result we get in example 6.1.1 ◀

Example 6.1.4. *A quiz consists of ten multiple choice questions. Four possible answers are provided for each question, and one of those four answers is correct. Suppose a student is totally unprepared and randomly chooses an answer for each question. Find the probability that the student answers*

(a) *Exactly three questions correctly.*

(b) Eight or more questions correctly.

(c) At least one question correctly.

Solution. *If we let "success" be "answering a question correctly", then $p = 1/4$ and $q = 1 - p = 3/4$. $n = 10$ since there are ten questions. Let X be the number of questions answered correctly.*

(a) The probability of answering three questions correctly is

$$\Pr(X = 3) = C(10,3)\left(\frac{1}{4}\right)^3 \left(\frac{3}{4}\right)^{10-3} \approx 0.2503$$

(b) "Answering eight or more questions correctly" consists of "answering 8, 9, or 10 questions correctly", and these events are mutually exclusive.

$$\Pr(X \geq 8) = \Pr(X = 8) + \Pr(X = 9) + \Pr(X = 10)$$
$$= C(10,8)\left(\frac{1}{4}\right)^8 \left(\frac{3}{4}\right)^2 + C(10,9)\left(\frac{1}{4}\right)^9 \left(\frac{3}{4}\right)^1 + C(10,10)\left(\frac{1}{4}\right)^{10} \left(\frac{3}{4}\right)^0$$
$$\approx 0.0004$$

(c) "Answering at least one question correctly" is the complement of "answering zero questions correctly".

$$\Pr(X \geq 1) = 1 - \Pr(X = 0) = 1 - C(10,0)\left(\frac{1}{4}\right)^0 \left(\frac{3}{4}\right)^{10} \approx 0.9437$$

The probability is quite high that the student will answer at least one question correctly :)◄

Before we move on to the next section, let's take a look at why the binomial distribution formula is the way it is. The number of different sequences that contain k successes in n Bernoulli trials is $C(n, k)$ because we have to choose k spots out of n to place these successes. The probability of each sequence is $p^k q^{n-k}$ because there are k successes and $n{-}k$ failures and they are independent, hence the formula $C(n, k) p^k q^{n-k}$ (math can be so precise, it is creepy, or beautifully creepy, or creepily beautiful—and by the way, congratulate yourself if you are reading this paragraph, because the majority of the people who need to read this book will miss this part. I just know. You are special!).

.:⋮ **EXERCISES 6.1** ⋮:.

1. Suppose an experiment has two possible outcomes x and y. If $\Pr(x) = 0.4$ then $\Pr(y) = $ ____ .

2. Suppose an experiment has three possible outcomes x, y, and z. If $\Pr(x) = 0.4$ and $\Pr(y) = $

0.5, then $\Pr(z) = $ ____.

3. Suppose an experiment has four possible outcomes, and we use the letter X to represent the outcome when the experiment is performed. X is called the _____ associated with the experiment.

4. You roll a fair three-sided die. The possible outcomes are $\{R, S, P\}$. Find the probability distribution of the random variable associated with the experiment. (Comment: there are fancy shaped three-sided dice you can find. The easiest way to make one is probably take a six-sided die and label two sides with one value, two other sides another value, and the remaining two sides yet another value.)

5. A box contains eight red and seven green numbered marbles. A sample of four marbles is randomly taken. Let X represent the number of red marbles in the sample and Y the number of green marbles in the sample.

   (a) Find the probability distribution of X.

   (b) Find the probability distribution of Y.

   (c) Explain why the probability distribution of Y is obvious once the probability distribution of X is found.

6. A robotic basketball player is to shoot three free throws. The probabilities that it makes the first, the second, and the third free throws are 0.7, 0.8, and 0.9, respectively. Let X be the number of free throws made. Suppose these three free throws are independent of each other, find the probability distribution of X.

7. A Bernoulli trial has _____ possible outcomes, one we call a _____ and the other a _____.

8. The binomial distribution with parameters n and p comes from counting the number of _____, each of which has probability _____, in a sequence of _____ independent Bernoulli trials.

9. A robot robocalls 10 telephone numbers. Suppose the probability that a robocall gets answered by a human is 0.25.

   (a) What is the probability that two of those ten calls are answered by a human?

   (b) What is the probability that two or fewer of the calls are answered by a human? $2, 1, 0$

   (c) What is the probability that at least one of the calls is answered by a human?

10. A robot robocalls 10 telephone numbers. Suppose the probability that a robocall gets answered by another robot that is meaner and nastier is 0.90.

(a) What is the probability that five or more of those calls are answered by a meaner, nastier robot?

(b) What is the probability that none of the calls are answered by a meaner, nastier robot?

11. A fair six-sided conventional die is rolled 10 times. The possible outcomes of each roll are $\{1,2,3,4,5,6\}$.

(a) Find the probability that the outcome 5 occurs twice.

(b) Find the probability that the outcome 5 occurs at least twice.

12. When the number of trials in the binomial distribution is very large, the calculation of some probabilities may become time consuming. For example, if X is a random variable that follows a binomial distribution with parameters 3000 and 0.7, and we want to find the probability $\Pr(X > 2100)$, then we will have to calculate

$$\Pr(X > 2100) = \Pr(X = 2101) + \cdots + \Pr(X = 3000)$$
$$= C(3000, 2101)0.7^{2101}0.3^{899} + \cdots + C(3000, 3000)0.7^{3000}0.3^{0}$$

There is a way to obtain an approximation, and it involves the *normal distribution*. Please do some research on this topic and write a report that includes: (a) the procedure, (b) situations in which this method applies, and (c) three different examples that show how this method works.

6.2 THE EXPECTED VALUE

Suppose an experiment has a finite number of outcomes, and the outcomes are real numbers (if they are not real numbers, we can always use real numbers to represent them, for example, 0 for heads and 1 for tails in a coin tossing experiment). Suppose the sample space of the experiment is $S = \{x_1, x_2, \ldots, x_k\}$, and the probability distribution of the random variable X associated with the experiment is $\Pr(X = x_i) = p_i$, $i = 1, 2, \ldots, k$. The *mean*, or *expected value*, of the random variable is given below. The common notation for the expected value is $E(X)$, or a single Greek letter μ.

The mean, or expected value, of a random variable X

$$\mu = E(X) = x_1 p_1 + x_2 p_2 + \cdots + x_k p_k \qquad (6.2.1)$$

What information does the expected value carry, though? This seemingly simple formula, I would argue, is the backbone of many, many modern business schemes.

Let's first examine the average, or to be mathematically correct, the *arithmetic average* (you may want to look up the term *geometric average*). This is the "everyday" average most of us refer to whenever we use the term "average". For example, if running back J.J. Smith had three carries of 5, 19, and 24 yards in a game, then he averaged 16 yards per carry because $(5 + 19 + 24)/3 = 16$.

Example 6.2.1. *There were 20 quizzes in my Math-575 class. The scores I obtained in those quizzes were 10, 10, 10, 9, 3, 9, 7, 7, 3, 10, 9, 7, 3, 3, 10, 10, 10, 9, 9, 3. What was my average quiz score?*

Solution. *It looks straightforward enough. We add all the numbers together and divide the sum by 20:*

$$\frac{10 + 10 + 10 + 9 + 3 + 9 + 7 + 7 + 3 + 10 + 9 + 7 + 3 + 3 + 10 + 10 + 10 + 9 + 9 + 3}{20} = 7.55$$

That looks good. But some of us are probably feeling a little unsatisfied. What if we need to find the average of 200 numbers? Adding too many numbers one by one is not only tedious, but also error-prone—we can easily miss one or two numbers here and there and double checking is... who wants to do double checking?

Progress comes from not being satisfied with the status quo. We look at the numbers again and notice that many of the numbers repeat. We can rearrange the numbers so scores of the same value are grouped together. Suppose we arrange the scores from small to large: 3, 3, 3, 3, 3, 7, 7, 7, 9, 9, 9, 9, 9, 10, 10, 10, 10, 10, 10, 10. And, instead of listing a number several times and then counting, we can just list a number and its count. This gives rise to a frequency distribution:

Score	Frequency
3	5
7	3
9	5
10	7

The frequency 5 associated with the score 3 indicates that the score 3 occurred 5 times. When we calculate the average, we will then multiply 3 by 5 to get the sum of the 5 quizzes that had a score of 3 each, instead of adding five 3's together. The average can be found by

$$\frac{3 \times 5 + 7 \times 3 + 9 \times 5 + 10 \times 7}{20} = 7.55$$

That looks better! But what if we multiply the numbers in the frequency column by 10? Since every score is repeated ten times as often, the total number of scores is now 200, and the average is

$$\frac{3 \times 50 + 7 \times 30 + 9 \times 50 + 10 \times 70}{200} = 7.55$$

It is still the same average! But why? Because the frequency-to-total ratios remain the same. How often does the number 3 occur? In the original case, it is 5 times out of 20: 5/20 = 1/4. In the modified case, it is 50 times out of 200: 50/200 = 1/4, same ratio. We will see the structure once we separate the terms in the numerator and keep the quiz scores in their original form:

$$\frac{3 \times 5 + 7 \times 3 + 9 \times 5 + 10 \times 7}{20} = \frac{3 \times 5}{20} + \frac{7 \times 3}{20} + \frac{9 \times 5}{20} + \frac{10 \times 7}{20}$$
$$= 3 \times \frac{5}{20} + 7 \times \frac{3}{20} + 9 \times \frac{5}{20} + 10 \times \frac{7}{20}$$

The numbers 5/20, 3/20, 5/20, and 7/20 are the relative frequencies associated with outcomes 3, 7, 9, and 10, respectively. If we randomly draw a score from the 20 scores, the probability of drawing a 3 is 5/20, so 5/20 is like the probability of the score 3. The form above has the structure of "sum of (outcome × probability)'s", and that is exactly the formula (6.2.1)◄

Example 6.2.2. *In a billing period of 30 days, my double-platinum-no-limit-presidential credit card carried a balance of $6000 for 10 days, $15000 for 5 days, $30000 for 8 days, and $50000 for 7 days. What was the average daily balance on the credit card over that billing period?*

Solution. *Sorting through the numbers, we see there are 10 $6000's, 5 $15000's, 8 $30000's and 7 $50000's. The frequency and relative frequency distributions are*

Balance	Frequency (number of days)	Relative frequency
6000	10	10/30
15000	5	5/30
30000	8	8/30
50000	7	7/30

The average daily balance is

$$\frac{6000 \times 10 + 15000 \times 5 + 30000 \times 8 + 50000 \times 7}{30} \approx 24166.67$$

Or if we use the relative frequencies

$$6000 \times \frac{10}{30} + 15000 \times \frac{5}{30} + 30000 \times \frac{8}{30} + 50000 \times \frac{7}{30} \approx 24166.67 ◄$$

Example 6.2.3. *You have completed 60 credit hours towards your degree. Of the 60 credit hours, 27 are A's, 21 are B's, and 12 are C's. If an A is worth 4 points, a B 3 points, and a C 2 points, what is your GPA?*

Solution. *By translating A, B and C to 4, 3 and 2, respectively, we obtain the following frequency and relative frequency distribution:*

Grade point	Frequency (number of credits)	Relative frequency
4	27	27/60
3	21	21/60
2	12	12/60

The grade point average (GPA) is then

$$\frac{4 \times 27 + 3 \times 21 + 2 \times 12}{60} = 3,25$$

Or if we use relative frequencies

$$4 \times \frac{27}{60} + 3 \times \frac{21}{60} + 2 \times \frac{12}{60} = 3,25 \blacktriangleleft$$

The last two examples above both follow the formula (6.2.1). That's the power of mathematical abstraction—one formula applies to many different situations.

The expected value of a random variable is the *long term average* of the values of the random variable. In other words, if we perform the experiment associated with a random variable over and over till forever, the average value of all the outcomes will approach the expected value of the random variable. For example, suppose we flip a fair coin over and over, and let the random variable X take on a value of 0 if the coin lands heads up and 1 if the coin lands tails up, then the expected value of X is $E(X) = 0 \times 0.5 + 1 \times 0.5 = 0.5$. In real life, we can't actually do an experiment forever. For a real life sequence of experiments to have an average that is close to the expected value, the experiment has to be performed "enough" times (I will take credit if you become a statistician someday after reading this sentence and wondering what "enough" means).

.:: **EXERCISES 6.2** ::.

1. The expected value of a random variable is the _____ average of the values of the random variable.

2. Suppose a student has received either A's or B's in all the classes she has taken, and that an A is equivalent to 4 grade points while a B is equivalent to 3. Is this information enough for us to conclude this student has a 3.5 Grade Point Average (GPA) because $(4 + 3)/2 = 3.5$? (By the way, never write $4 + 3/2 = 3.5$ because $4 + 3/2 = 4 + 1.5 = 5.5$. Sloppy writing is bad for math.)

3. The probability distribution of a random variable is given below. Find its expected value.

Outcome	Probability
2	0.2
4	0.3
6	0.4
8	0.1

4. The probability distribution of a random variable is given below. Find its expected value.

Outcome	Probability
−3	0.15
0	0.10
1	0.45
2	0.30

5. In this remote mountain cabin there is no wireless connection to the outside world. It is snowing heavily outside so you and your brother decide to play tic-tac-toe all day. Whoever wins a game wins a dollar from the other person. Suppose you win 55% of the time. What is the expected value of the game from your point of view? From your brother's point of view?

6. You sell your hand-made bracelets on the Internet. For each bracelet sold, you make a profit of $10. For each bracelet that's returned for a replacement, you make a profit of $5. For each bracelet that is returned for a full refund, you lose $3. What is your expected profit per bracelet if 5% of your bracelets are returned for a replacement and 3% are returned for a full refund?

7. You mix three alcohol solutions together to create a new solution. The three solutions you use are: 5 gallons of 60% alcohol, 4 gallons of 45% alcohol, and 3 gallons of 30% alcohol. What is the percent of alcohol in the final solution?

8. There are 12 100-point quizzes in an online class you are taking. If you score 60% on 5 of the quizzes, 45% on 4 of the quizzes, and 30% on 3 of the quizzes, what is your average quiz score?

9. A dental hygienist spends one hour with a patient 20% of the time, 45 minutes 50% of the time, and 30 minutes 30% of the time. What is the average amount of time the dental hygienist spends with one patient?

10. Determine whether this statement is true or false: If the expected value of a random variable is 7, then 7 is the most likely outcome of the experiment associated with the random variable.

11. A tutoring service guarantees that anyone who takes its driving test preparation course will pass the test, or it will refund 110% of the tuition to the student. Suppose the tutoring

service charges $2000 per student for the course, and 70% of its students pass the driving test. What is the average revenue per student for the tutoring service?

6.3 DECISION MAKING AND BUSINESS SCHEMES

Even though abstraction can be powerful, it has to come from experiencing enough real life situations. We are going to go all real life now, and hopefully make a connection between abstraction and application.

Example 6.3.1. *You were a natural leader a long time ago in a village far, far away. People liked to surround you and follow your ideas. One day you had this brilliant idea: What if I can help people and make a profit for myself at the same time? That would be like killing two stones with one bird. So you told your people: "Listen, last month Mr. Obi-Two Cannot Be got sick and had to see a doctor, but the medical bill was 100 shovels of olives and Mr. Obi-Two Cannot Be just couldn't afford it and he ended up escaping the village and now we don't have someone to tell us wonderful stories after supper. What a loss for all of us. But think about it, if each of us had given Mr. Obi-Two Cannot Be just one shovel of olives, that would have been more than enough to pay for his medical bill. What does one shovel of olives mean to one family? Well, let me just say you wouldn't even notice it if I didn't tell you, it is like you clip one string of hair from your horse's mane, what does the horse care?! And for that one string of hair we could have saved Mr. Obi-Two Cannot Be from leaving us! Who wouldn't want to do that?" The village people all roared in unison: "Only the heartless would do that, but definitely not us!" And you continued: "The problem is, it is hard to collect one shovel of olives from each family on short notice, and we all know the doctor charges outrageous interest if we don't pay the bill immediately. So I am proposing that each of you give me one shovel of olives every month. I will keep them in my very safe personal vault. If any of you need to pay a large bill in an emergency, you just need to tell me and I will take the olives from the vault to pay your bill. You will be saved and everyone else will be a hero! How does that sound?" And people roared again: "That's the most brilliant idea we have never heard! Long live our dearest greatest leaderest!" And from then on that was the law of the village. But you didn't tell people that emergencies didn't happen very often, and when one happened, you only needed on average 50 shovels of olives to take care of it—the village had 500 families!*

This was the beginning of the modern insurance industry (according to me)◀

Example 6.3.2. *You go to your first class meeting of math-132—Finite Mathematics. There are 30 students in the class. The funny looking professor asks every student in class to give him $2 for a chance to win $50 at the end of the class. What he will do is he will randomly pick a student in the class and give $50 to this student. Since there are 30 students in the class, this professor will net a profit of $10. Before you cry "highway robbery" and "crooked professor", please realize this is exactly how a state sponsored lottery works. The state collects a certain amount of money* *from ticket sales, redistributes a certain percentage of the ticket sales as winning prizes to a few people, keeps the rest as profit, and maybe uses some of it to improve some areas of concern. If I were the professor, I would give $2 of my $10 profit to a charity and tell you that you are in a win-win situation: if you win $50 then obviously you win; if you don't win $50 then your money goes to a charity for the right cause—yes, you pay $2, but it is for a good cause. Come on, that's a win, right? You've just become a better person! Well, that's only partially true because I didn't tell you a big part of your money goes into my pocket!◄*

Example 6.3.3. *I was run out of the village because I did something that really displeased the village leader. I wandered west and came to this desert town situated between two civilizations. I decided to settle down here. How could I make a living, though? Well, opening a casino in the desert seemed like a good idea. And so I opened a casino. There was only one game people played in my casino: You pick a number from 1 to 6, and pay a clam to the casino hostess. The hostess rolls a fair six-sided die with numbers 1, 2, 3, 4, 5, 6 on its six faces, one number on each face. If the number on the top face of the die happens to be the one you picked, you collect five clams from the hostess; otherwise you lose your clam. What is the expected value of the game from my (the casino's) perspective?*

Solution. *Since the die is fair, the probability that the top face has any number from 1 to 6 is 1/6. So with probability 1/6 the casino loses 4 clams (this is the part a lot of gamblers don't catch: if you win, you win only 4 clams because you pay the casino 1 clam to play the game and then you get 5 clams back, a net winning of 4 clams), and with probability 5/6 the casino wins 1 clam. The probability distribution of the winnings for the casino is*

Winnings	Probability
+1 🐚	5/6
−4 🐚's	1/6

Expected value = $(1)(5/6) + (-4)(1/6) = 1/6 \approx 0.17$ clams. This means that in the long run, when a lot of games have been played, the casino can expect to collect 0.17 clams per game. It doesn't matter who the players are, the casino simply treats all gamblers as one giant gambler. So if 50 people come in one day and they collectively roll a die 10,000 times, then the casino's revenue is expected to be $0.17 \times 10000 = 1700$ clams◄

Example 6.3.4. *Your brother sells teddy bears at $20 apiece. You want to sell a one year warranty for the teddy bears. Based on his years of experience, your brother knows that about 10% of bears get messed up in the first year by their owners. A warranty will require you to replace a messed up bear with a new one, which costs $20 because your brother won't give you a discount and he has exclusive rights to all teddy bears under the sky. What is the minimum amount per bear you should charge for a one year warranty in order not to lose money?*

Solution. *Suppose you set the warranty price at x dollars. A teddy bear goes unwell with probability 0.10, and in that case you will lose $(20-x)$ dollars because you need to spend 20 dollars to buy a* *new bear but you already collected x dollars from the customer, your net loss is (20–x) dollars. With probability 0.90 the bear will be well and sound and in that case you pocket the x dollars from selling the warranty. If we use P to represent your profit per bear, then the distribution of P is*

P	Probability
x	0.90
$-(20-x)$	0.10

The expected value of P is

$$E(P) = 0.90x - 0.10(20-x) = x - 2$$

For you not to lose money, E(P) must be positive, so you will have to set the warranty price above $2 per bear.

In real life, you may have other expenses to consider. So in order to be profitable you may need to set the price much higher than $2. Obviously you can't set the price too high, either, because otherwise people would say "to the fourth dimension with your warranty, I will just buy a new bear if this one doesn't bathe itself"◄

The next example is also real life, but the math involved can be found in section 6.1, example 6.1.2. The connection between these two examples is not so obvious, though. So in that sense, real life' is sometimes harder than abstraction.

The game and the numbers in the following example are fabricated, but that of course is not important. The model is used in all lottery games, including the Keno games.

Example 6.3.5. *To play the Super Six Harmony Color Lottery, a player picks six numbers from the numbers 1 to 36, and pays $2 to the government to have the six numbers printed on a nicely designed ticket. Later an absolutely fair and righteous robot appointed by the government randomly picks six balls from a drum that contains 36 balls numbered 1 to 36. If the six numbers picked by the robot totally match the six numbers picked by the player, the player wins the grand prize of $1,000,000. If only five numbers among the six picked by the robot are among the six numbers picked by the player, the player wins the second prize of $10,000. If only four numbers among the six picked by the robot are among the six numbers picked by the player, the player wins a consolation prize of $100. Let X be the player's winnings.*

   (a) Find the probability distribution of X.
   (b) Find the expected value of X.

Solution. *When the player picks 6 numbers out of 36, the 36 numbers are divided into two groups: 6 picked by the player and 30 not picked by the player. The total number of ways this can be accomplished is $C(36,6) = 1947792$. For ease of reference, let's call the 6 numbers picked by the* *player group A and the other* 30 *numbers group B. The fair and righteous robot then picks* 6 *numbers of its own. If these* 6 *numbers are all in group A, the player wins the grand prize. If* 5 *of these* 6 *numbers are in group A and* 1 *is in group B, the player wins the second prize. If* 4 *of these* 6 *numbers are in group A and* 2 *are in group B, the player wins the consolation prize.*

(a) *The probability distribution of X is as follows*

X	Probability
999998	$C(6,6)C(30,0)/C(36,6) \approx 0.000000513$
9998	$C(6,5)C(30,1)/C(36,6) \approx 0.000092412$
98	$C(6,4)C(30,2)/C(36,6) \approx 0.003349947$
−2	$1 - sum\ of\ the\ three\ above \approx 0.996557128$

(b) *The expected value of X is*

$$E(X) = 999998 \times 0.000000513 + 9998 \times 0.000092412$$
$$+ 98 \times 0.003349947 + (-2) \times 0.996557128 \approx -0.23$$

This means that in the long run, for every $2 *you spend on a ticket, you are expected to say goodbye to* 23 *cents. A few players will get rich, the government doesn't care. If the citizens collectively spend* $20,000,000 *on lottery tickets, the government collects* $2,300,000. *Some of the money will be put to good use such as buying the latest gadget for kindergarten kids who can then use the device to cyberbully each other during school hours, some will be used to pay for the wining and dining of some government officials◄*

Example 6.3.6. *It is known that one of the* 100 *barrels of whisky is contaminated because the person who contaminated it confessed. The problem is this person can't remember exactly which barrel is contaminated. An emergency meeting is called and someone suggests: "We will just have to take some of the good stuff from each barrel, one at a time, and test it, until we find the contaminated barrel." Everyone says: "Yeah, that sucks, but it beats throwing all the barrels away. They are the single malt kind that's worth a lot of money." And so they proceed to do just that. If they are really lucky, the first barrel they test is contaminated and they don't need to do any more testing. If they are really, really unlucky, the first* 99 *barrels are uncontaminated and they will have tested* 99 *barrels before they find the contaminated one. And, let's assume the testing is quite expensive and time consuming so they would like to perform as few tests as possible.*

As we can see here, the number of tests needed to find the contaminated barrel ranges from 1 *to* 99 *(we don't need the* 100th *test because the* 99th *test either catches the contaminated barrel, or it implicates the* 100th *barrel), and there is no way we can know ahead of time how many tests need to be run. If we let X be the number of tests needed to find the contaminated barrel, we can find the expected value of X and that would be the average number of tests needed. To do that, we first need to find the probability distribution of X. This experiment has* 99 *outcomes, and this can seem a little intimidating. What do we do when we are facing an intimidating problem? We try anyway until we succeed or until we fail. If we don't try, then we just fail. So try we will. Here we go:*

X = 1 if the first barrel is contaminated. This has a probability 1/100 of happening because there are 100 barrels and we happen to pick the one that's contaminated.

X = 2 if the first barrel is not contaminated and the second one is. This has a probability (99/100)(1/99) of happening because there are 99 uncontaminated barrels out of 100 in the beginning so the probability of picking an uncontaminated one is (99/100). Once we find out the first one is not contaminated, we pick a second one but at this point there are only 99 barrels left so the probability of picking the contaminated one is (1/99). Let's do the easy part now: (99/100)(1/99) = 1/100. OK, we got the number... oh, wait a minute, this is exactly the same as the probability that the first one is contaminated. Is this a coincidence, or is there something going on?

X = 3 if the first two are not contaminated and the third one is. The probability of this happening is (99/100)(98/99)(1/98) = 1/100, 1/100 again! We can analyze a few more but pretty soon we are convinced that the probabilities are all equal to 1/100 for X = 1, 2, 3, . . . , 98.

X = 99 is a little different because this happens when the first 98 barrels are not contaminated. The probability that this happens is (99/100)(98/99)(97/98) · · · (3/4)(2/3) = 2/100.

The probability distribution of X is

X	*Probability*
1	1/100
2	1/100
3	1/100
⋮	⋮
98	1/100
99	2/100

The expected value of X is

$$E(X) = 1 \times \frac{1}{100} + 2 \times \frac{1}{100} + 3 \times \frac{1}{100} + \cdots + 98 \times \frac{1}{100} + 99 \times \frac{2}{100}$$
$$= \frac{1}{100}(1 + 2 + 3 + \cdots + 98) + \frac{198}{100} = \frac{1}{100} \cdot \frac{98(1+98)}{2} + \frac{198}{100} = 50.49$$

The sum 1 + 2 + . . . + 98 is found using formula (1.1.2), our almost forgotten old friend from chapter 1. Like a loyal, good old friend, it comes to help when it is needed◄

Return to the result we have just obtained. With this straightforward testing scheme, we will have to test 50.5 times on average before we find the contaminated barrel. Can we improve? Not always, but in this case, yes! You may want to take a dance break before you go on. It surely has been hard work.

Our next method can be described as cool, as in Pink Panther—the Coolest Feline in History kind of cool. Even though it involves no random variables, it is related to the geometric sequence in chapter 1, and anytime we can bring seemingly unrelated things together, it is super cool.

Example 6.3.7. —*Another method to find the contaminated whisky barrel.*

Let's take one bottle of the good stuff from each barrel, and label them. Bring in two large tubes. (1) Divide the 100 bottles into two groups, 50 bottles in each group. Let's call them group A and group B. Pour a little bit from each bottle in group A into tube I, and pour a little bit from each bottle in group B into tube II. Now test tube I. If tube I is contaminated, the contaminated barrel is in group A. If tube I is not contaminated, the contaminated barrel is in group B. Let's assume tube I is contaminated, the 50 barrels that belong to group B are now exonerated. Can you sense the beauty of this method? One test eliminates half of the population! (2) Now we divide the 50 bottles in group A into two new groups, 25 bottles in each group. Repeat the first step and eliminate 25 barrels. (3) Divide the remaining 25 bottles into two groups, 13 in one and 12 in another. One more test and assume the 12-bottle group is exonerated. (4) Divide the remaining 13 bottles into a 7-bottle group and a 6-bottle group. Assume the 6-bottle group is exonerated after the test. (5) Divide 7 into 4 and 3. Assume 3 is exonerated. (6) Divide 4 into 2 and 2, eliminate one of them. (7) Divide 2 into 1 and 1, one final test and we catch the bad guy!

Total number of tests = 7. Guaranteed! How much is saved? 51–7 = 44, we have saved 44 unnecessary tests over the first method, that's 86% savings! Let's have a shot. Cheers! 🥃🥃

But why seven tests? How are the numbers 7 and 100 even related? Let's see. Every test we run eliminates half of the population, and we keep going until there are only two elements left in the population. At that point, one more test takes care of business. Now if we go backwards, before we run the final test, there are two barrels left, the step before that there should be four or fewer barrels, and then the step before that there should be eight or fewer barrels. 2, 4, 8, ..., it is a geometric sequence: 2, 2^2, 2^3, ... If we go six steps back, we are at a population size of 2^6 = 64, which is not greater than the original population of 100, but 2^7 = 128 > 100, that means we need 7 tests to reduce the original population down to one element. Aha! If the population is P, then the number of tests we need to run is the smallest number n such that 2^n > P. For example, if there are 200 barrels of whisky and one is contaminated, how many tests do we need to run to find the contaminated barrel? Surprisingly, we need only 8 tests even though the number of barrels *doubled, because* $2^7 = 128 < 200$, *so* 7 *tests won't do. But* $2^8 = 256 > 200$, *so* 8 *tests will do.*

Let's crack open one barrel and celebrate, shall we? ◄

Example 6.3.8. *The Emperor's spokesperson is a robot named OnePlusOneEqualsToo. It is programmed in such a way that* 90% *of statements it makes are false. Suppose OnePlusOneEqualsToo made* 8 *statements in the past hour. Let X be the number of true statements in those* 8 *statements.*

    (a) *Find the probability distribution of X.*
    (b) *Find the expected value of X.*

Solution. *This is a binomial distribution. Let "success" be "telling a true statement", we have* $p = 0.10$ *and* $n = 8$.

    (a) *Use formula (6.1.1), we have* $\Pr(X = k) = C(8, k)(0.1)^k(0.9)^{8-k}$, *where* $k = 0, 1, 2, \ldots, 8$. *The probability distribution of X is calculated below.*

X	Probability
0	0.43046721
1	0.38263752
2	0.14880348
3	0.03306744
4	0.00459270
5	0.00040824
6	0.00002268
7	0.00000072
8	0.00000001

    (b) *The expected value of X is found by formula (6.2.1),*

$$E(X) = 0 \times 0.43046721 + 1 \times 0.38263752 + 2 \times 0.14880348$$
$$+ 3 \times 0.03306744 + 4 \times 0.00459270 + 5 \times 0.00040824$$
$$+ 6 \times 0.00002268 + 7 \times 0.00000072 + 8 \times 0.00000001 = 0.8 ◄$$

Remark. *In the last example, the robot tells the truth* 10% *of the time. So out of* 8 *statements, it makes sense that the number of true statements will be* 10% *of* 8, *or* 0.8, *as we have calculated. The proof takes some clever algebraic manipulation and the reader is not required to fully understand it. We will nevertheless present the proof after we have stated the formula below so the curious types can have a pure mathematics experience*◄

The expected value of a binomial random variable with parameters n and p

$$E(X) = np \qquad\qquad (6.3.1)$$

Proof. From (6.1.1) and (6.2.1), we see that

$$E(X) = \sum_{k=0}^{n} kC(n,k)p^k q^{n-k} = \sum_{k=0}^{n} k\frac{n!}{k!(n-k)!}p^k q^{n-k} = \sum_{k=1}^{n} \frac{n!}{(k-1)!(n-k)!}p^k q^{n-k}$$

$$= \sum_{k=0}^{n-1} \frac{n!}{k!(n-1-k)!}p^{k+1} q^{n-1-k} = np\sum_{k=0}^{n-1} \frac{(n-1)!}{k!((n-1)-k)!}p^k q^{(n-1)-k} = np \blacktriangleleft$$

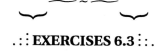

.:: **EXERCISES 6.3** ::.

1. You sell your homemade handbags for a living. You also want to sell a two-year replace-ment warranty on your handbags. Suppose it costs you $30 to replace a handbag. The percentages of the handbags that are returned for replacement once, twice, and three times within two years are 5%, 2%, and 1%, respectively. What is the minimum price at which you can sell your warranty without losing money?

2. Your anthropology class grade consists 20% of homework, 20% of quizzes, 30% of a big project, and 30% of the final exam. Entering the final exam you have earned 95% on homework, 82% on quizzes, and 89% on the project. What percentage should you earn on the final exam to have a 90% overall percentage in the class?

3. A quiz consists of 10 multiple choice questions, with 4 possible answers for each question and one of them being the correct answer. Suppose someone randomly selects one answer for each question. Let X be the number of correct answers selected.

   (a) Find the probability distribution of X.

   (b) Find the expected number of correct answers.

4. You bring your pet robot to a country fair for pet robots. There is a nerdy host robot at a booth that asks a pet robot player three challenging questions. It takes 100 Unicoins to play one game. Suppose the probability of answering a question correctly is 0.10, and the player earns 300 Unicoins for every question answered correctly.

   (a) Let X be the number of questions answered correctly. Find the probability distribu-tion of X.

   (b) Find the expected value of the game for your pet robot.

5. Same game as above, but this time a lot more challenging (for you, the reader that is, not your pet robot—it doesn't care) with a change of the rules. Here are the rules: If the player answers the first question correctly, it will be offered a prize of 500 Unicoins to leave the

game, or no prize to continue to the second question. If the player chooses to continue, and answers the second question correctly, it will be offered a prize of 7100 Unicoins to leave the game, or no prize to continue to the third question. If the player chooses to try the third question, the prize will be 100100 Unicoins for the correct answer. Assume with probability 0.5 the player will choose to continue when offered the chance. Find the expected value of the game for your pet robot. (Hint: A tree diagram would be a good tool for analyzing the game. From the tree you can more clearly see all the possible outcomes. Calculate the probabilities carefully and come up with the probability distribution. From there the expected value can be found.)

6. Five freshmen and seven sophomores run for four committee seats on the Social Committee. They are all good friends and all well qualified so they decide to have a robot make the selection for them. The robot randomly picks 4 names out of the 12 names given to it. Let X be the number of freshmen selected by the robot.

   (a) Find the probability distribution of X.

   (b) Find the expected value of X.

7. Five freshmen, four sophomores and three juniors run for four committee seats on the Social Committee. They are all good friends and all well qualified so they decide to have a robot make the selection for them. The robot randomly picks 4 names out of the 12 names given to it. Let X be the number of freshmen selected by the robot.

   (a) Find the probability distribution of X.

   (b) Find the expected value of X.

6.4 THE VARIANCE AND THE STANDARD DEVIATION

The expected value is a measure of *central tendency*. Two random variables can have the same expected value yet behave quite differently. For example, in the National Football League running back A has run for 40, 50, 60, 70, 80, 90, 100, 110, 120, and 130 yards over the first ten games of the season, his average is $(40 + 50 + 60 + 70 + 80 + 90 + 100 + 110 + 120 + 130)/10 = 85$ yards/game. Running back B has run for 80, 85, 85, 85, 85, 85, 85, 85, 85, and 90 yards over the first ten games of the season, his average is $(80 + 85 + 85 + 85 + 85 + 85 + 85 + 85 + 85 + 90)/10 = 85$ yards/game. We can see from these numbers that B is more like a robot, he just goes roughly the same distance every game, as if he stops only when his battery runs dry. A on the other hand is more exciting and has ups and downs from game to game. We can see the difference between these two samples because each sample contains only 10 numbers and the numbers are pretty easy to read. Imagine if we have two samples each containing 1000 numbers and we still want to know which sample has numbers that are more spread out, we will have trouble doing it with the naked eye. We need some measure to show us the spread.

The standard deviation is one of the statistics measures of spread (or dispersion). As the term measure of *spread* suggests, we want to know how concentrated or spread out the numbers in a group are. One way to measure that is by taking the average of the distances from every number in the group to the mean of the group. For example, let's say group A contains numbers 3, 4, 7, 9, 12, its mean is $\mu = (3 + 4 + 7 + 9 + 12)/5 = 7$. The sum of the distances from every number to the mean is $|3 - 7| + |4 - 7| + |7 - 7| + |9 - 7| + |12 - 7| = 14$, the absolute values are necessary so we don't get both positive and negative distances that cancel each other out. The average distance is $14/5 = 2.8$. This number gives us a good idea how far away from the mean the numbers in the group are, on average. It is a good measure, but it is rarely used because the absolute value operation is not pleasant to deal with when it comes to performing various mathematical manipulations.

The exponent 2 is used instead. The squaring operation has the effect of turning all real numbers to nonnegative just like the absolute value operation does, but it is much easier to deal with in a lot of other ways, among them the operation of differentiation in calculus. So if we take the average of the square of distances from every number to the mean, we get

$$\frac{(3 - 7)^2 + (4 - 7)^2 + (7 - 7)^2 + (9 - 7)^2 + (12 - 7)^2}{5} = 10.8$$

This quantity is called the variance. If the numbers in the group come with a unit, say "miles", then the variance will have the unit "square miles". We can recover the original unit by taking the square root of the variance: $\sqrt{10.8 \text{ square miles}} \approx 3.29$ miles, and this quantity is called the standard deviation, denoted by the Greek letter σ.

If we have a larger group of numbers, and some numbers in the group repeat, such as the following group of 25 numbers:

$$2, 2, 2, 3, 4, 4, 5, 5, 5, 5, 5, 5, 6, 6, 7, 7, 7, 7, 8, 8, 8, 8, 8, 9$$

The mean and standard deviation are straightforward to find—the reader is encouraged to find those right now with paper and pencil and the help of a simple calculator. The mean is $\mu = 5.64$ and the standard deviation is $\sigma \approx 2.02$. We have seen how the mean is calculated when repeating numbers present in a group in section 6.2. Following the same reasoning we arrive at the following formulas.

Suppose the sample space of an experiment is $S = \{x_1, x_2, \ldots, x_k\}$, and the probability distribution of the random variable associated with the experiment is $\Pr(X = x_i) = p_i$, $i = 1, 2, \ldots, k$. Let μ be the expected value of X. The variance and the standard deviation of the random variable are defined as

> The variance of a random variable
>
> $$\sigma^2 = (x_1 - \mu)^2 p_1 + (x_1 2 - \mu)^2 p_2 + \cdots + (x_k - \mu)^2 p_k \qquad (6.4.1)$$

> The standard deviation of a random variable
>
> $$\sigma = \sqrt{(x_1 - \mu)^2 p_1 + (x_1 2 - \mu)^2 p_2 + \cdots + (x_k - \mu)^2 p_k} \qquad (6.4.2)$$

Given two comparable groups of numbers, the one with the larger variance or standard deviation has numbers that are more spread out. We will use the running backs A and B in our introductory example to demonstrate.

Example 6.4.1. *Group A consists of the following ten numbers* 40, 50, 60, 70, 80, 90, 100, 110, 120, 130. *Group B consists of the following ten numbers:* 80, 85, 85, 85, 85, 85, 85, 85, 85, 90. *Suppose each number in each group occurs with probability 1/10.*

  (a) *Find the standard deviations for A and B.*

  (b) *Based on the standard deviations, which group of numbers is considered more consistent?*

Solution.

  (a) *For group A:*

$$\mu_A = \frac{1}{10} \cdot 40 + \frac{1}{10} \cdot 50 + \frac{1}{10} \cdot 60 + \frac{1}{10} \cdot 70 + \frac{1}{10} \cdot 80$$
$$+ \frac{1}{10} \cdot 90 + \frac{1}{10} \cdot 100 + \frac{1}{10} \cdot 110 + \frac{1}{10} \cdot 120 + \frac{1}{10} \cdot 130 = 85$$

$$\sigma_A = \sqrt{\begin{array}{l} \frac{1}{10}(40-85)^2 + \frac{1}{10}(50-85)^2 + \frac{1}{10}(60-85)^2 + \frac{1}{10}(70-85)^2 \\[6pt] + \frac{1}{10}(80-85)^2 + \frac{1}{10}(90-85)^2 + \frac{1}{10}(100-85)^2 + \frac{1}{10}(1100-85)^2 \\[6pt] + \frac{1}{10}(120-85)^2 + \frac{1}{10}(130-85)^2 \end{array}} \approx 28.72$$

  For group B:

$$\mu_B = \frac{1}{10} \cdot 80 + \frac{1}{10} \cdot 85 + \frac{1}{10} \cdot 85 + \frac{1}{10} \cdot 85 + \frac{1}{10} \cdot 85$$
$$+ \frac{1}{10} \cdot 85 + \frac{1}{10} \cdot 85 + \frac{1}{10} \cdot 85 + \frac{1}{10} \cdot 85 + \frac{1}{10} \cdot 90 = 85$$

$$\sigma_B = \sqrt{\begin{array}{l} \frac{1}{10}(80-85)^2 + \frac{1}{10}(85-85)^2 + \frac{1}{10}(85-85)^2 + \frac{1}{10}(85-85)^2 \\ + \frac{1}{10}(85-85)^2 + \frac{1}{10}(85-85)^2 + \frac{1}{10}(85-85)^2 + \frac{1}{10}(85-85)^2 \\ + \frac{1}{10}(85-85)^2 + \frac{1}{10}(90-85)^2 \end{array}} \approx 2.24$$

(b) *Since the standard deviation for group B is much smaller than that for group A, the numbers in group B are more consistent because they don't deviate from the mean as much as the numbers in group A* ◄

Example 6.4.2. *Let's revisit example 6.3.5 in section 6.3. Find the standard deviation of X.*

Since only Prob → no Fraction we just mult (pay't don't divide mean) [handwritten note]

X	Probability
999998	$C(6,6)C(30,0)/C(36,6) \approx 0.000000513$
9998	$C(6,5)C(30,1)/C(36,6) \approx 0.000092412$
98	$C(6,4)C(30,2)/C(36,6) \approx 0.003349947$
−2	$1 - $ *sum of the three above* ≈ 0.996557128

Solution. *We have already calculated the mean:* $\mu = -0.23$. *The standard deviation is*

$$\sigma = \sqrt{\begin{array}{l} (999998-(-0.23))^2 \times 0.000000513 + (9998-(-0.23))^2 \times 0.000092412 \\ + (98-(-0.23))^2 \times 0.003349947 + ((-2)-(-0.23))^2 \times 0.996557128 \end{array}} \approx 722.68$$

That is a large number for a standard deviation. What it tells us is that the distribution is wildly turbulent, and of course we kind of know it already: a few people get a lot of money, and almost everyone loses money—that's wild, for the winners! ◄

We close this section with the standard deviation formula for the binomial distribution. A proof is provided. But like the proof of the expected value formula for the binomial distribution, it is sigma-notation intensive and is meant to be read by people who enjoy the formula manipulation aspect of mathematics.

The standard deviation of a binomial random variable with parameters n and p (where $q = 1 - p$)

$$\sigma = \sqrt{npq} \tag{6.4.3}$$

Proof. First we introduce an alternate formula for the variance of a random variable. From

(6.4.1), we have

$$\sigma^2 = \sum_{i=1}^{k} (x_i - \mu)^2 p_i = \sum_{i=1}^{k} (x_i^2 - 2x_i\mu + \mu^2) p_i$$

$$= \sum_{i=1}^{k} x_i^2 p_i - 2\mu \sum_{i=1}^{k} x_i p_i + \mu^2 \sum_{i=1}^{k} p_i = \sum_{i=1}^{k} x_i^2 p_i - 2\mu\mu + \mu^2 = E(X^2) - \mu^2$$

With this new formula, we first find $E(X^2)$ for a binomial distribution with parameters n and p:

$$E(X^2) = \sum_{k=0}^{n} \frac{n!}{k!(n-k)!} p^k q^{n-k} k^2 = \sum_{k=1}^{n} \frac{n!}{(k-1)!(n-k)!} p^k q^{n-k} k$$

$$= \sum_{j=0}^{n-1} \frac{n!}{j!(n-1-j)!} p^{j+1} q^{n-1-j}(j+1) = np \sum_{j=0}^{n-1} \frac{(n-1)!}{j!(n-1-j)!} p^j q^{n-1-j}(j+1)$$

$$= np \sum_{j=0}^{n-1} C(n-1,j) p^j q^{n-1-j}(j+1)$$

$$= np \left[\sum_{j=0}^{n-1} C(n-1,j) p^j q^{n-1-j} j + \sum_{j=0}^{n-1} C(n-1,j) p^j q^{n-1-j} \right]$$

$$= np \left[(n-1)p + 1 \right] = n^2 p^2 + npq$$

Now

$$\sigma^2 = E(X^2) - \mu^2 = n^2 p^2 + npq - (np)^2 = npq$$

and

$$\sigma = \sqrt{npq} \blacktriangleleft$$

You may want to put (6.1.1), (6.3.1), and (6.4.3) together to form a summary of the binomial distribution. I could have copy-and-pasted the three formulas here for you of course, but believe me, by looking for them and writing them down yourself, you will have a much more permanent impression of the formulas.

.:: **EXERCISES 6.4** ::.

1. The expected value is a statistical measure of _____. The variance is a statistical measure of _____.

2. If we take a probability distribution, and change the highest outcome value to a larger number, the variance will (choose one answer) <u>increase</u> / <u>decrease</u>.

3. Find the variance and standard deviation for the following probability distribution.

Outcome	Probability
0	0.25
1	0.25
2	0.25
3	0.25

(handwritten note: Can't assume that # is n. We don't have n.)

4. Find the variance and standard deviation for the following probability distribution.

Outcome	Probability
−1	0.90
0	0.05
1	0.03
2	0.02

5. Five freshmen and seven sophomores run for four committee seats on the Social Committee. A robot makes the selection for them by randomly picking 4 names out of the 12 names given to it. Let X be the number of freshmen selected by the robot. Find the variance and standard deviation of X.

6. A fair two-sided coin is flipped 100 times. Let W be the total number of heads in 100 flips. Find the expected value, the variance, and the standard deviation of W.

7. A fair six-sided die has the numbers 1, 2, 3, 4, 5, 6 on its six faces, one on each face. The die is rolled 180 times. Let Y be the total number of 6's in 180 rolls. Find the expected value, the variance, and the standard deviation of Y.

8. A scamming robot makes 1500 threatening calls each day. Suppose the probability of successfully making the recipient believe the threat is 0.01. Find the expected value, the variance, and the standard deviation of the number of successful threats made.

6.5   OTHER MEASURES IN STATISTICS

We are not trying to be encompassing, we are trying to be intriguing and inspiring. Some statistics measures have been left out because the author doesn't have much to say—they are either known to everyone, or can be searched and understood in a few minutes. Some other statistics measures are left out because the author doesn't know much about them—his deep apologies here.

One measure that's most worth mentioning is the *median*. Like the mean, the median is a measure of central tendency. The median is the number that divides a group of numbers into two equal halves: one half consists of numbers that are less than or equal to the median, and the other numbers greater than or equal to the median. That sounds easy enough but there are minor complications and some rules need to be established and agreed upon.

We first arrange the numbers in ascending order. If a group of numbers consists of an odd number of numbers, then there is one number in the group that is at the center of the group and this number is the median. For example, if Jack is 6 ft 2 in tall, Tom is 5 ft 8 in tall, and Sylvester is 5 ft 10 in tall, then Sylvester's height is the median in this group of three numbers: 5 ft 8 in, 5 ft 10 in, 6 ft 2 in.

If a group of numbers consists of an even number of numbers, then there is no one single number in the group that is at the center. In this case we take the arithmetic average of the two numbers that are the closest to the center as the median.

Example 6.5.1. *Find the median for each of the following two groups of numbers.*
Group A: 10, 9, 4, 7, 8, 5, 5, 6.
Group B: 10, 9, 4, 7, 8, 5, 5, 6, 2.

Solution. *Let's rearrange the numbers so they are in ascending order.*

Group A: 4, 5, 5, 6, 7, 8, 9, 10. *There are* 8 *numbers, the two at the center are* 6 *and* 7. *The median is* $(6 + 7)/2 = 6.5$.

Group B: 2, 4, 5, 5, 6, 7, 8, 9, 10. *There are* 9 *numbers, the one at the center is* 6. *The median is* 6 ◄

Both the median and the mean have been frequently used as a representative of a population. A lot of times they are quite close. For example, for group *A* in the last example, the mean is 6.75, which is pretty close to the median 6.5. The natural question is then why do we need both? Are there times when one measure is more appropriate than the other?

The median is easier to calculate because it depends on the relative positions of the numbers, and its calculation involves at most two numbers. The median is close to the mean when the numbers in the group are evenly spread out.

The mean, as we can see from its formula, takes all the numbers into account. The mean can be affected by a few extreme (very large or very small compared to the population) numbers in the group. For example, the group 2, 3, 5, 7, 9 has mean 5.2 and median 5. If we add one large number 100 to the group: 2, 3, 5, 7, 9, 100, then the mean becomes 21, way different from the mean 5.2 of the first group, but the median is 6, still relatively close to the median of the first group. For this reason, sometimes people take out the extreme numbers in a group before they calculate the mean. For example, if a sporting event involves scoring by several judges, sometimes the highest and the lowest scores are taken out before the mean is calculated. Or if we want to calculate the average per capita income of a country but the income of the Royal family of that country accounts for half of the country's total income, then it may be a good idea to not include the Royal family's income in the calculation because that one super rich family makes the regular folks look richer than they actual are.

Another central tendency measure is called the *mode*. The mode is the most popular number in a group. For example, in the group of numbers 3, 4, 5, 5, 5, 5, 6, 6, 7, the mode is 5 because 5 has the highest frequency: four. The concept seems simple yet it can encounter some unusual situations. For example, there may be more than one number that occurs most often, such as in the group: 1, 2, 2, 2, 3, 3, 4, 4, 4, 5, 6, both 2 and 4 occur three times. Or when every number occurs exactly once such as: 7, 8, 9, 10. If you are interested, do some research so you can know everything about the mode.

In addition to the standard deviation, there are other measures of spread. The *range* is simply the difference between the largest number (maximum) and the smallest number (minimum) in a group. The *inter-quartile range* is the difference between the third quartile and the first quartile. The *first quartile* is the median of the lower half of the group, and the *third quartile* is the median of the upper half of the group.

The newly introduced definitions are demonstrated using the following group of numbers.

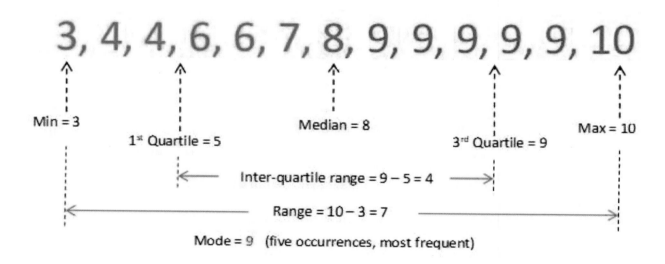

.:: **EXERCISES 6.5** ::.

1. Fill in each blank with one of the following words to make the statement most correct: *Significant, Insignificant, Immediate, Less, More.*

(a) Adding a number that is much greater than every number in a group of numbers to that group (you are welcome to provide suggestions that will make this sentence less annoying) will have a _Significant_ impact on the value of the mean for the group.

(b) Adding a number that is much smaller than every number in a group of numbers to that group will have much ___more___ impact on the value of the mean than on the median for the group.

2. Find the mean and median for each group of numbers.

   (a) 4,4,5,5,5,5,6,6,6. Median=5   46/9=5.11=mean

   (b) 4,4,5,5,5,5,6,6,6,120. Median=5   166/10= 16.6 =mean

   (c) Say something intelligent about the two results above.

3. The NoFoodShouldBeWasted restaurant is owned and operated by 20 similar-minded young people who are both owners and workers. They share the profits and workload equally. In describing the typical average income per worker of the restaurant, which of the following measures are appropriate? (Check all that apply.)

   (A) The mean   (B) The median   (C) The mode   (D) The range

4. The TheBestEverNotEvenClose restaurant is owned by a billionaire who does the cooking himself and hires 20 workers. The owner makes 3 million dollars per year but no worker makes more than $30,000 per year. Which of the following measures would be appropriate to describe the average wages of a typical worker in that restaurant?

   (A) The median of everyone's income, including the owner's.

   (B) The mean of everyone's income, including the owner's.

   (C) The mean of everyone's income, not including the owner's.

   (D) The standard deviation, excluding the owner's number.

5. Given two groups of numbers

$$A: 2,3,5,5,5,5,5,5,5,7 \qquad Mode=5$$
$$B: 1,2,3,4,5,5,6,7,8,9 \qquad Mode=5$$

   (a) Find the mode of each group.

   (b) The mode is a better representative for which group of numbers? Why?   A = 5,5's + smaller spread

6. Given the following group of numbers, find the median, the min, the max, the 1st quartile, the 3rd quartile, the range, and the inter-quartile range.   Med: 106.5

$$100, 102, 102, 105, 106 \mid 107, 108, 109, 110, 110$$

   Q_1 Q_3

Median = 106.5   Mode=102 + 110

106+107 /2 = 106.5

Min = 100

Max = 110

Range = 110-100 = 10

Q_1 = 102

Q_3 = 109

IQR = 109-102 = 7

Chapter 7

Some Interesting Topics and Case Studies

This chapter presents some topics that may involve deeper or broader applications of some of the material we have studied, or simple yet totally surprising applications of some of the material we have studied.

7.1 MARKOV CHAINS—WHEN MATRICES TEAM UP WITH PROBABILITIES

We like it whenever we must solve a problem using several tools because we get the satisfaction of solving a complex problem and also that of showing off the tools in our possession. It is like how we clean a dirty dish: detergent, sponge, running water, and a sink that is not clogged. Dirty dish in, clean dish out, mission accomplished! Very impressive stuff!

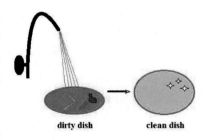

dirty dish   clean dish

A *Markov chain* is a sequence of random variables such that the distribution of any random variable in the sequence depends solely on the distribution of the random variable preceding it. In other words, if the sequence of random variables $X_0, X_1, X_2, \ldots$ form a Markov chain, then the distribution of X_1 depends only on the distribution of X_0, the distribution of X_2 depends only on the distribution of X_1, the distribution of X_3 depends only on the distribution of X_2, and so on. We call the distribution of X_0 the *initial distribution*.

Example 7.1.1. *There is a fixed number of robotic fish in a tank. A fish is either red or green on any given day, but they can change color from day to day. Suppose the probability that a red fish will remain red the next day is 0.7, and the probability that a green fish will remain green the next*

151

day is 0.6.

(a) *If on a given day* 50% *of the fish are red and* 50% *are green, what percent of the fish will be red, and what percent will be green the next day?*

(b) *What about two days later?*

(c) *Three days later?*

(d) *Will the percentage of red fish and that of green fish eventually become constants?*

Solution. *We will use tree diagrams to solve the problem first, and develop a better method that utilizes matrices later.*

(a) *First we draw a tree. Use R for "red", G for "green", and RG for "red today and green the day after", etc.*

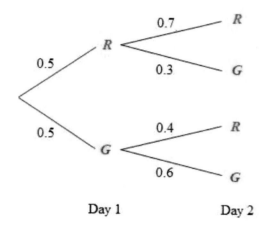

From the tree diagram we can see that after one day, the probability of a fish being red is

$$\Pr(red\ one\ day\ later) = \Pr(RR) + \Pr(GR) = (0.5)(0.7) + (0.5)(0.4) = 0.55$$

And the probability of a fish being green is

$$\Pr(green\ one\ day\ later) = \Pr(RG) + \Pr(GG) = (0.5)(0.3) + (0.5)(0.6) = 0.45$$

(b) *Another day, the tree grows another layer—this is when we start to wonder if there is a better way. Because this method is going to be very labor intensive if we want to know what happens after ten days. Well for now we have the tree so let's soldier on:*

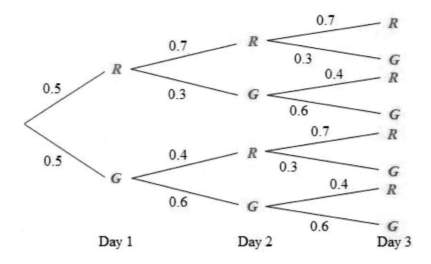

$$\Pr(red\ two\ days\ later) = \Pr(RRR) + \Pr(RGR) + \Pr(GRR) + \Pr(GGR)$$
$$=(0.5)(0.7)(0.7) + (0.5)(0.3)(0.4) + (0.5)(0.4)(0.7) + (0.5)(0.6)(0.4) = 0.565$$

$$\Pr(green\ two\ days\ later) = \Pr(RRG) + \Pr(RGG) + \Pr(GRG) + \Pr(GGG)$$
$$=(0.5)(0.7)(0.3) + (0.5)(0.3)(0.6) + (0.5)(0.4)(0.3) + (0.5)(0.6)(0.6) = 0.435$$

Now we are thinking: Do we have to do the next part? Well, yes. But let's see if we can find a better way. After one day, the percentages of the colors change, but the probabilities of going from one color to another are still the same, red to red 0.7, red to green 0.3, green to red 0.4, and green to green 0.6. The computations we have just done can be accomplished by using the first tree diagram with the initial distribution of (R, G) changed from (0.5, 0.5) to the distribution after one day, (0.55, 0.45):

$$\Pr(red\ two\ days\ later) = (0.55)(0.7) + (0.45)(0.4) = 0.565$$

$$\Pr(green\ two\ days\ later) = (0.55)(0.3) + (0.45)(0.6) = 0.435$$

(c) To find the distribution three days later, we repeat the computations above using the distribution after two days, (0.565, 0.435), as the initial distribution:

$$\Pr(red\ three\ days\ later) = (0.565)(0.7) + (0.435)(0.4) = 0.5695$$

$$\Pr(green\ three\ days\ later) = (0.565)(0.3) + (0.435)(0.6) = 0.4305$$

(d) Our confidence is growing and overflowing, we say "bring on 10 days later, or 100 days later if you like..." Oh, maybe not. To get to 100 days, we need to pass the first 99 days, and that's still some daunting mileage even though the road is straight and flat. We need a better tool before we can go to war.

All the computations we have performed above involve four constants: 0.7, 0.3, 0.4, and 0.6. If we examine the steps closely, we see a pattern:

$$\Pr(R \text{ next day}) = \Pr(R \text{ today})(0.7) + \Pr(G \text{ today})(0.4)$$

$$\Pr(G \text{ next day}) = \Pr(R \text{ today})(0.3) + \Pr(G \text{ today})(0.6)$$

And this is when magic happens. The two equations above can be written as a matrix equation:

$$\begin{bmatrix} \Pr(R \text{ next day}) \\ \Pr(G \text{ next day}) \end{bmatrix} = \begin{bmatrix} 0.7 & 0.4 \\ 0.3 & 0.6 \end{bmatrix} \begin{bmatrix} \Pr(R \text{ today}) \\ \Pr(G \text{ today}) \end{bmatrix}$$

If we use x_0 and y_0 to represent the initial distribution of red and green fish, and x_i and y_i the distribution of red and green fish after i days, then the matrix above can be written as

$$\begin{bmatrix} x_{i+1} \\ y_{i+1} \end{bmatrix} = \begin{bmatrix} 0.7 & 0.4 \\ 0.3 & 0.6 \end{bmatrix} \begin{bmatrix} x_i \\ y_i \end{bmatrix} \quad \text{where } i = 0, 1, 2, \dots, \text{ and } \begin{bmatrix} x_0 \\ y_0 \end{bmatrix} = \begin{bmatrix} 0.5 \\ 0.5 \end{bmatrix}$$

The square matrix above is called the transition matrix (for a Markov chain). The form of the equation suggests a "geometric sequence of matrices". The distribution of red and green fish after 2 days is

$$\begin{bmatrix} x_2 \\ y_2 \end{bmatrix} = \begin{bmatrix} 0.7 & 0.4 \\ 0.3 & 0.6 \end{bmatrix} \begin{bmatrix} x_1 \\ y_1 \end{bmatrix} = \begin{bmatrix} 0.7 & 0.4 \\ 0.3 & 0.6 \end{bmatrix} \begin{bmatrix} 0.7 & 0.4 \\ 0.3 & 0.6 \end{bmatrix} \begin{bmatrix} x_0 \\ y_0 \end{bmatrix} = \begin{bmatrix} 0.7 & 0.4 \\ 0.3 & 0.6 \end{bmatrix}^2 \begin{bmatrix} 0.5 \\ 0.5 \end{bmatrix}$$

In general, the red and green fish population after n days is

$$\begin{bmatrix} x_n \\ y_n \end{bmatrix} = \begin{bmatrix} 0.7 & 0.4 \\ 0.3 & 0.6 \end{bmatrix}^n \begin{bmatrix} 0.5 \\ 0.5 \end{bmatrix}$$

For example, after 10 days the distribution is

$$\begin{bmatrix} x_{10} \\ y_{10} \end{bmatrix} = \begin{bmatrix} 0.7 & 0.4 \\ 0.3 & 0.6 \end{bmatrix}^{10} \begin{bmatrix} 0.5 \\ 0.5 \end{bmatrix} \approx \begin{bmatrix} 0.571428 \\ 0.428572 \end{bmatrix}$$

And after 100 days the distribution is

$$\begin{bmatrix} x_{100} \\ y_{100} \end{bmatrix} = \begin{bmatrix} 0.7 & 0.4 \\ 0.3 & 0.6 \end{bmatrix}^{100} \begin{bmatrix} 0.5 \\ 0.5 \end{bmatrix} \approx \begin{bmatrix} 0.571429 \\ 0.428571 \end{bmatrix}$$

We see that the distribution after 10 days and the distribution after 100 days are almost identical. We will not get into the details here because the details can interfere with the appreciation of the big picture. We just want to have a taste of it, be awed by it, and have a glimpse of a light by which we can venture into the wonderland if we so choose. We would like to know if the percentages of red fish and green fish will settle into constants as time goes on. Based on the distributions after 10 days and after 100 days, we would like to think the answer is yes and the percentages are approximately 57% for red fish and 43% for green fish. But can we find the exact percentages? It turns out the answer is yes and finding the percentages is not too difficult. From the equation

$$\begin{bmatrix} x_{i+1} \\ y_{i+1} \end{bmatrix} = \begin{bmatrix} 0.7 & 0.4 \\ 0.3 & 0.6 \end{bmatrix} \begin{bmatrix} x_i \\ y_i \end{bmatrix}$$

we see that if the percentages of red fish and green fish settle into constants x and y as time goes on (i.e., as i approaches ∞), then

$$\begin{bmatrix} x \\ y \end{bmatrix} = \begin{bmatrix} 0.7 & 0.4 \\ 0.3 & 0.6 \end{bmatrix} \begin{bmatrix} x \\ y \end{bmatrix}$$

This is equivalent to the system

$$\begin{cases} 0.7x & + & 0.4y & = & x \\ 0.3x & + & 0.6y & = & y \end{cases}$$

The two equations in the system both simplify to $0.3x - 0.4y = 0$. But since x and y make up the entire fish population, $x + y = 1$. Adding this equation to the system we get

$$\begin{cases} 0.3x & - & 0.4y & = & 0 \\ x & + & y & = & 1 \end{cases}$$

The solution to the system is

$$\begin{bmatrix} x \\ y \end{bmatrix} = \begin{bmatrix} 4/7 \\ 3/7 \end{bmatrix} \approx \begin{bmatrix} 0.57 \\ 0.43 \end{bmatrix}$$

In other words, as time passes, there will be 4 red and 3 green fish for every 7 fish in the tank.

It is worth noting that the initial distribution has no effect on the eventual distribution. The transition probabilities alone determine the eventual distribution.

It is also worth noting that the example can be applied to humans. Just change "red fish" to "registered as a member of party A" and "green fish" to "registered as a member of party B".

And, seriously, this example could have been listed as an application in (a) robotic science, (b) aquatic fish science, and (c) political science, if there were an Index of Applications in this book◄

The example can be generalized to three colors or more. The transition matrices will be larger but all the operations remain the same.

::: **EXERCISES 7.1** :::

1. In a remote town far, far away, two auto repair shops operate right next to each other. Shop A goes by the philosophy of cultivating loyal customers. 95% of A's customers will return to A for the next service, and 5% will go to B for the next service. Shop B on the other hand operates under the Olden rule of "If I can take advantage of you today, I definitely won't wait until tomorrow." As a consequence 80% of B's customers will go to A for the next service, and 20% will return to B for the next service. Suppose initially 50% of all customers go to A and 50% go to B. As time goes on, what percent of customers in the town will be serviced by A and what percent will be serviced by B?

2. There are two Chinese restaurants in a small town. The Wongs offers the best Chinese food. Its competitor, the Qongs, cooks lousy, salty, and ugly Chinese food. 90% of Wongs' customers return to Wongs for their next meal, 10% will try Qongs just to see how bad it is. 50% of Qongs' customers will go to the Wongs for their next meal, 50% will go back to Qongs because some people they don't like are going over to Wongs and they just want to show these people they don't care. In the long run, what percent of the town's Chinese restaurant goers will eat at Wongs?

3. Three big fish ponds are connected by narrow canals. By the King's order (the king is a huge catfish by the way), a fish can move from one pond to another at most once a day. 70% of the fish in pond A stay in pond A the next day, 20% will move to pond B, and 10% will move to pond C. 60% of the fish in pond B stay in pond B the next day, 25% will move to pond A and 15% will move to pond C. 80% of the fish in pond C stay in pond C the next day, 5% will move to pond A and 15% will move to pond B. In the long run, what percent of the fish will be in each pond every day? (Hint: If you use x, y, z to represent the long-term distribution, add $x + y + z = 1$ to your system of equations.)

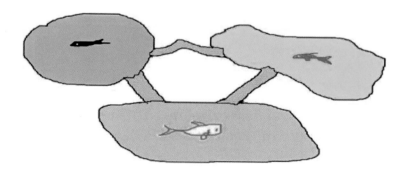

7.2 HOW MANY ROUTES ARE THERE FROM POINT A TO POINT B?

Another way to interpret combination and partition is through the study of routes from one point to another on a grid. Let's start with a two-dimensional grid.

Example 7.2.1. *Refer to the graph below. Suppose one can move only north (up) or east (to the right) along the edges. How many routes are there from point A to point B?*

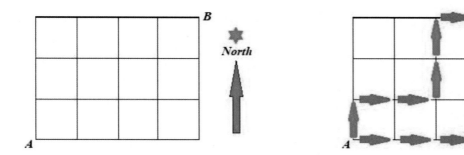

Solution. *Use an N to represent moving one block north and an E east. Two possible routes are: EEEENNN and NEENNEE. There are many more of course but it is not a good idea to try to list all the possible routes because there are too many of them and we may not know when we have found all of them. If we look at the two routes we have had, we notice that in both sequences, there are four E's and three N's. If we come up with a couple more routes we soon realize this has to be true for all the routes because one has to move four blocks east and three blocks north, in any order, in order to go from A to B. The problem is reduced to filling seven spots with four E's and three N's. This couldn't be easier, we have done it in chapter 4. The answer is*

$$C(7,4)C(3,3) = C(7,4) = 35$$

Another potentially very messy problem is reduced to a simple combination!◄

Example 7.2.2. *Refer to the graph below. Suppose one can move only north (up) or east (to the right) along the edges.*

   (a) *How many routes are there from point A to point B if one has to stop at point C?*
   (b) *If one randomly chooses a route from A to B, what is the probability that one passes point C?*

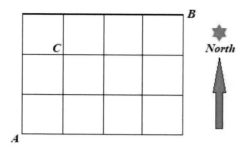

Solution. *We are feeling relaxed and confident now that we have seen some examples.*

(a) *To go from A to C one has to pass three blocks, one of them E and two of them N, so the number of routes is $C(3,1) = 3$. To go from C to B one has to pass four blocks, three of them E and one of them N, the number of routes is $C(4,3) = 4$. The total number of routes from A to B passing through C is $3 \times 4 = 12$.*

(b) *Putting the answer from above and the last example together, the probability that a randomly chosen route passes through C = 12/35* ◄

Example 7.2.3. *In the future in a city the streets are 3-D and people can walk not just horizontally but also vertically (ants have been doing that already). Suppose one can move only east (to the right), north (to the back), or up (towards sky) along the edges. Refer to the diagram below. How many routes are there from point A to point B?*

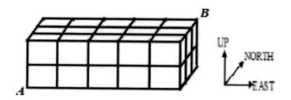

Solution. *This takes a little bit of 3-D visualization. The reasoning is similar to that of example 7.2.1. In order to go from A to B one must cover 10 blocks: two blocks UP, three blocks NORTH, and five blocks EAST. The total number of routes is therefore*

$$C(10,2)C(8,3)C(5,5) = 2520$$

Or we can use partition

$$\binom{10}{2,3,5} = \frac{10!}{2!3!5!} = 2520$$

Let's stop and appreciate the power of mathematics here. 2520 different routes! Imagine if we actually have to list all the different routes and count them! What if we just add one block to each direction? It will be 60060 routes from point A to point B ◄

:∶: **EXERCISES 7.2** :∶:

1. A lives 8 street blocks from B (see map below). Suppose all streets are clean and safe. In how many ways can A walk to B's place to study for an exam together if he walks only north (up) or east (to the right)?

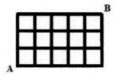

2. Refer to the given figure. Suppose one can move only to the left or down along the edges.

   (a) How many routes are there from point A to point B?

   (b) How many routes are there from point A to point B that pass through point C?

   (c) How many routes are there from point A to point B that pass through point C and point D?

   (d) If a route is randomly chosen, what is the probability that this route passes through C?

   (e) If a route is randomly chosen, what is the probability that this route passes through C and D?

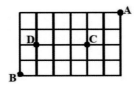

3. Prince W's army is descending upon the evil Dragon Lord. The Lord is holed up in a castle in the southeast corner of city Z. W's army is camped in the north-west corner of city X (see map below). At dawn, W's army will attack with full force, crush stronghold Y first, and then seize the Lord's castle. Based on the map, in how many different ways can W's army reach the Dragon Lord's castle if the army will march only south / east / southeast-ish?

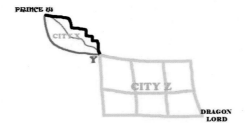

4. (This one is a little harder) A lives 8 street blocks from B (see figure below). In how many ways can B walk to A's place to study for an exam together if she walks only south (down) or west (to the left)?

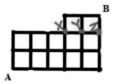

5. A scaffolding is made of eight hollow cubes. Suppose an ant can move only to the right, to the back, or down along the edges. How many routes are there from point A to point B?

7.3   SET THEORY AND LOGIC

If someone claims that it was recorded that the great sprinter Usain Bolt ran a 100-meter dash in 8 seconds on September 7, 2017 in Los Angeles, we can verify that the claim is false.

It is easier to judge whether one single, simple declarative statement is true or false. If we put several such statements together using connective words such as *AND* and *OR* to form a compound statement, it becomes more difficult to judge whether the compound statement is true or false. That's where logic comes in. Logic is a subject by itself and there are rules you follow. Here we want to show how some logic problems can be solved with the help of sets.

First, let's get familiar with the language of logic.

1. A *statement* is a declarative sentence that is either true or false.
2. Suppose p and q are two statements. The expression $p \wedge q$ stands for "statement p AND statement q".
3. The expression $p \vee q$ stands for "statement p OR statement q".
4. The expression $\neg p$ stands for "NOT statement p", or "NEGATION of statement p".
5. The expression $p \rightarrow q$ stands for "IF p THEN q".

Let p and q be two sets.

1. We use an element x to determine whether a statement is true or false. If $x \in p$ then p is true; if $x \notin p$ then p is false.

2. $p \wedge q$ is equivalent to the intersection $p \cap q$.
3. $p \vee q$ is equivalent to the union $p \cup q$.
4. $\neg p$ is equivalent to the complement p'.
5. $p \rightarrow q$ is equivalent to p being a subset of q. Because if p is a subset of q, then $x \in p$ implies $x \in q$. For example, if $p = \{$all mammals$\}$ and $q = \{$all animals$\}$, then $p \subseteq q$, and $p \rightarrow q$ as a consequence. To verify, let x be a cat, then $x \in p$, therefore $x \in q$.

Here are the elementary truth tables and their corresponding set operation equivalents. The set operations, especially when aided by Venn diagrams, are easy to comprehend. The truth tables, on the other hand, are usually memorized when one learns logic. There is nothing wrong with memorization, but it is also nice when we can cleverly explain something with diagrams.

If p	Then $\neg p$	Set Operations
T	F	If $x \in p$, then $x \notin p'$
F	T	If $x \notin p$, then $x \in p'$

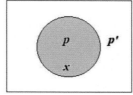 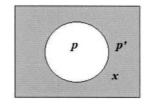

If		Then	Set
p	q	$p \vee q$	Operations
T	T	T	If $x \in p$ and $x \in q$, then $x \in p \cup q$
T	F	T	If $x \in p$ and $x \notin q$, then $x \in p \cup q$
F	T	T	If $x \notin p$ and $x \in q$, then $x \in p \cup q$
F	F	F	If $x \notin p$ and $x \notin q$, then $x \notin p \cup q$

If		Then	Set
p	q	$p \wedge q$	Operations
T	T	T	If $x \in p$ and $x \in q$, then $x \in p \cap q$
T	F	F	If $x \in p$ and $x \notin q$, then $x \notin p \cap q$
F	T	F	If $x \notin p$ and $x \in q$, then $x \notin p \cap q$
F	F	F	If $x \notin p$ and $x \notin q$, then $x \notin p \cap q$

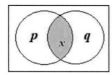 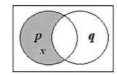 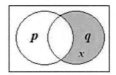 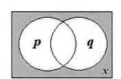

This set approach to logic also applies to more complicated compound statements. If you have learned logic the traditional way, you can try using this new approach to solve some old problems and see if this new approach is better in some ways and worse in some other ways than the traditional approach. If you have never learned logic before, it is comforting to know that mastering the basic set operations already makes you a decent logician.

And one more truth table before we end the section:

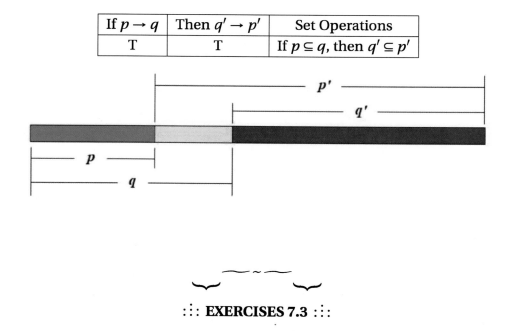

If $p \rightarrow q$	Then $q' \rightarrow p'$	Set Operations
T	T	If $p \subseteq q$, then $q' \subseteq p'$

⌣‿ ～ ‿⌣

:∴: **EXERCISES 7.3** :∴:

1. You are driving on a freeway and a beautiful truck goes right by you. The truck has two gasoline tanks and the engine can draw gasoline from either tank.

   (a) Fill in the blank with AND or OR to make the following statement a true statement: The truck runs as long as tank 1 _____ tank 2 is not empty.

   (b) Use a Venn diagram to show the corresponding set operation. Use one circle to represent the situation "tank 1 is empty" and another circle the situation "tank 2 is empty".

2. Given a true statement "If A works, then B works". Use a Venn diagram to show that the statement "If B does not work, then A does not work" is also a true statement.

3. Use a Venn diagram to point out the fallacy in the following argument: "If a person robs a bank, this person is a bad guy. The police just arrested two bad guys, these two must have robbed some bank."

4. Use a Venn diagram to point out the fallacy in the following statement: "It is known that every time it rains, some weeds grow in my yard afterwards. It has not rained yet, so no weeds will grow in my yard."

7.4 A PICTURE IS A PICTURE BUT NOT THE SAME PICTURE

We have learned some matrix operations in chapter 3, and we have seen how an image can be stored as a matrix. Suppose M is a matrix that stores an image. I want to send M to you electronically but the image is top secret so I would like to scramble it before I send it to you in case someone intercepts the image. I can use a matrix A to do the scrambling. Instead of sending you M, I send you AM, then send you A^{-1} separately. You use A^{-1} to unscramble the image: $A^{-1}(AM) = M$. (Just some inspiring ideas here. Real life scrambling and unscrambling are a little bit more sophisticated.)

Example 7.4.1. *Below is a simple image along with its associated matrix M (I know they look familiar). In the matrix a 0 stands for NO COLOR, a 1 GREEN, a 2 YELLOW, and a 3 RED. Suppose I choose the the matrix A as the scrambling matrix, then AM records the scrambled face:*

$$M = \begin{bmatrix} 0 & 0 & 0 & 0 & 0 & 0 & 0 & 0 \\ 0 & 0 & 0 & 0 & 0 & 0 & 0 & 0 \\ 0 & 1 & 1 & 0 & 0 & 2 & 2 & 0 \\ 0 & 1 & 1 & 0 & 0 & 2 & 2 & 0 \\ 0 & 0 & 0 & 0 & 0 & 0 & 0 & 0 \\ 0 & 0 & 0 & 0 & 0 & 0 & 0 & 0 \\ 0 & 0 & 3 & 3 & 3 & 3 & 0 & 0 \\ 0 & 0 & 0 & 0 & 0 & 0 & 0 & 0 \end{bmatrix}$$

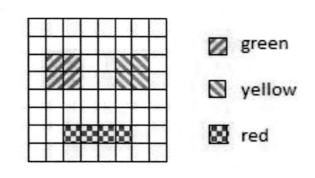

green
yellow
red

$$A = \begin{bmatrix} 1 & 0 & 0 & 0 & 0 & 0 & 0 & 0 \\ 0 & 0 & 0 & 0 & 0 & 1 & 0 & 0 \\ 0 & 1 & 0 & 0 & 0 & 0 & 0 & 0 \\ 0 & 0 & 1 & 0 & 0 & 0 & 0 & 0 \\ 0 & 0 & 0 & 0 & 0 & 0 & 1 & 0 \\ 0 & 0 & 0 & 0 & 0 & 0 & 0 & 1 \\ 0 & 0 & 0 & 0 & 1 & 0 & 0 & 0 \\ 0 & 0 & 0 & 1 & 0 & 0 & 0 & 0 \end{bmatrix}$$

$$AM = \begin{bmatrix} 0 & 0 & 0 & 0 & 0 & 0 & 0 & 0 \\ 0 & 0 & 0 & 0 & 0 & 0 & 0 & 0 \\ 0 & 0 & 0 & 0 & 0 & 0 & 0 & 0 \\ 0 & 1 & 1 & 0 & 0 & 2 & 2 & 0 \\ 0 & 0 & 3 & 3 & 3 & 3 & 0 & 0 \\ 0 & 0 & 0 & 0 & 0 & 0 & 0 & 0 \\ 0 & 0 & 0 & 0 & 0 & 0 & 0 & 0 \\ 0 & 1 & 1 & 0 & 0 & 2 & 2 & 0 \end{bmatrix}$$

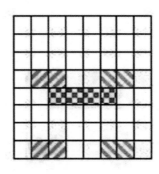

The inverse of A is

$$A^{-1} = \begin{bmatrix} 1 & 0 & 0 & 0 & 0 & 0 & 0 & 0 \\ 0 & 0 & 1 & 0 & 0 & 1 & 0 & 0 \\ 0 & 0 & 0 & 1 & 0 & 0 & 0 & 0 \\ 0 & 0 & 0 & 0 & 0 & 0 & 0 & 1 \\ 0 & 0 & 0 & 0 & 0 & 0 & 1 & 0 \\ 0 & 1 & 0 & 0 & 0 & 0 & 0 & 0 \\ 0 & 0 & 0 & 0 & 1 & 0 & 0 & 0 \\ 0 & 0 & 0 & 0 & 0 & 1 & 0 & 0 \end{bmatrix}$$

With that, you unscramble the image back to its original form by performing the multiplication $A^{-1}(AM) = M$ ◄

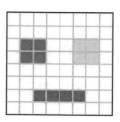

$$\underbrace{} \overset{\frown}{\sim} \underbrace{}$$

∴∴ **EXERCISES 7.4** ∴∴

1. Given the following "face"

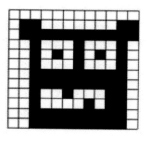

   (a) Use a matrix *M*, with 1 representing black and 0 representing white, to record the "face".

(b) Use the following matrix A to find the product AM.

$$A = \begin{bmatrix} 0 & 0 & 0 & 0 & 0 & 1 & 0 & 0 & 0 & 0 & 0 & 0 \\ 1 & 0 & 0 & 0 & 0 & 0 & 0 & 0 & 0 & 0 & 0 & 0 \\ 0 & 0 & 1 & 0 & 0 & 0 & 0 & 0 & 0 & 0 & 0 & 0 \\ 0 & 0 & 0 & 0 & 0 & 0 & 0 & 1 & 0 & 0 & 0 & 0 \\ 0 & 0 & 0 & 0 & 0 & 0 & 1 & 0 & 0 & 0 & 0 & 0 \\ 0 & 1 & 0 & 0 & 0 & 0 & 0 & 0 & 0 & 0 & 0 & 0 \\ 0 & 0 & 0 & 0 & 0 & 0 & 0 & 0 & 1 & 0 & 0 & 0 \\ 0 & 0 & 0 & 1 & 0 & 0 & 0 & 0 & 0 & 0 & 0 & 0 \\ 0 & 0 & 0 & 0 & 0 & 0 & 0 & 0 & 0 & 0 & 1 & 0 \\ 0 & 0 & 0 & 0 & 0 & 0 & 0 & 0 & 0 & 0 & 0 & 1 \\ 0 & 0 & 0 & 0 & 0 & 0 & 0 & 0 & 0 & 1 & 0 & 0 \\ 0 & 0 & 0 & 0 & 1 & 0 & 0 & 0 & 0 & 0 & 0 & 0 \end{bmatrix}$$

(c) Convert AM back to a "face". This is the scrambled "face".

(d) Use the matrix A^{-1} given below to find $A^{-1}AM$. Verify that this is indeed the matrix M.

$$A^{-1} = \begin{bmatrix} 0 & 1 & 0 & 0 & 0 & 0 & 0 & 0 & 0 & 0 & 0 & 0 \\ 0 & 0 & 0 & 0 & 0 & 1 & 0 & 0 & 0 & 0 & 0 & 0 \\ 0 & 0 & 1 & 0 & 0 & 0 & 0 & 0 & 0 & 0 & 0 & 0 \\ 0 & 0 & 0 & 0 & 0 & 0 & 0 & 1 & 0 & 0 & 0 & 0 \\ 0 & 0 & 0 & 0 & 0 & 0 & 0 & 0 & 0 & 0 & 0 & 1 \\ 1 & 0 & 0 & 0 & 0 & 0 & 0 & 0 & 0 & 0 & 0 & 0 \\ 0 & 0 & 0 & 0 & 1 & 0 & 0 & 0 & 0 & 0 & 0 & 0 \\ 0 & 0 & 0 & 1 & 0 & 0 & 0 & 0 & 0 & 0 & 0 & 0 \\ 0 & 0 & 0 & 0 & 0 & 0 & 1 & 0 & 0 & 0 & 0 & 0 \\ 0 & 0 & 0 & 0 & 0 & 0 & 0 & 0 & 0 & 0 & 1 & 0 \\ 0 & 0 & 0 & 0 & 0 & 0 & 0 & 0 & 1 & 0 & 0 & 0 \\ 0 & 0 & 0 & 0 & 0 & 0 & 0 & 0 & 0 & 1 & 0 & 0 \end{bmatrix}$$

7.5 WHAT'S VARIANCE GOT TO DO WITH IT?

As we have discussed before, the concept of the game of lottery is simple: The host takes a little bit of money from each individual in a large population, keeps a certain percentage of the money collected and returns the rest to a few individuals. For example, suppose the king of seven kingdoms has 300 million people under his rule. Every week the king asks every citizen to pay him $2 to purchase a lottery ticket. 5 citizens will get 100 million dollars each and they become instant celebrities and are envied by everybody else. It is an exciting event every week and people are always looking forward to it. It keeps the citizens preoccupied with something so they don't have time to see the many faults the king has. The winners are on all social media with big, happy grins from ear to ear. Every citizen thinks "next week it could be me!"

Scenario I: Every one of the 300 million people spends $2 for a ticket, that's 600 million dollars. The king gives 500 million dollars to five people, that leaves him with 100 million dollars to spare. Let's say he spends 20 million dollars on logistics, that means he still pockets 80 million dollars, EVERY WEEK! Good deal, right? Who would have thought ruling seven kingdoms could be so simple?

Scenario II: The king can also collect 100 million dollars if he simply asks every citizen to pay him 1/3 of a dollar every week. But that would be taxation and people really hate taxation. They might overthrow the king if he did that.

What is the difference between scenario I and scenario II if the king gets the same amount of money in either case? Excitement! Scenario I is exciting, scenario II is boring! And how is "exciting" defined in mathematics?

Suppose you were a citizen. In scenario I, your probability of winning $(100000000–2) is $5/300000000 = 1/60000000$, your probability of losing $2 is $59999999/60000000$. Your expected value of winnings is

$$99999998 \cdot \frac{1}{60000000} + (-2) \cdot \frac{59999999}{60000000} = -\frac{1}{3}$$

The variance of your winnings is

$$\left(99999998 + \frac{1}{3}\right)^2 \cdot \frac{1}{60000000} + \left(-2 + \frac{1}{3}\right)^2 \cdot \frac{59999999}{60000000} \approx 166666663.9$$

In scenario II, you lose 1/3 of a dollar with probability 1, your expected value of winnings is $(-1/3) \cdot 1 = -\$1/3$, and the variance of your winnings is $(-1/3 - (-1/3))^2 \cdot 1 = 0$, that's ZERO! You just turn in your 1/3 of a dollar every week and nothing happens after that! That's the definition of boring.

Anyway, or anyways, the main difference between the two scenarios is the variance. A large variance creates excitement, and people need excitement to carry on. How much money is spent on 10 minutes of fireworks in New Year's Eve in Las Vegas? I don't know but I would think it is enough money to buy food that can feed many hungry people for very many days so why do people still spend millions of dollars every year on fireworks and leave some hungry people hungrier? Because people need to get excited so they can work harder and generate more economic output that will in turn feed more hungry people. It is not a total waste of money, you see. But of course I have been totally off the course... this is a math book. Sorry about that.

Gambling (also called "gaming" if you are drunk and drop the letters b and l) is the simplest application of the expected value and the variance. Suppose you operate a casino, and you want to create a new game. The first thing you do is make sure the expected value is in your favor, and then you adjust the variance to make the game reasonably exciting so people want to play. Finally you use some of your profit to give the patrons free drinks and other stuff to convince them they are getting their money's worth so they come back for more. Viva Las Vegas!

Scenario II is as boring as you can get—a zero variance! No wonder communism doesn't work because communism tries to make people more uniform and that's boring... oh, am I off course again? Sorry, signing off now...

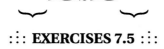

:⋮: **EXERCISES 7.5** :⋮:

1. If the variance of a random variable is 0, the random variable can be described as (choose one most appropriate word) exciting / boring / fair.

2. Two basketball players average about the same number of points per game. If we calculate the variances of their point distributions and find that the variance of one player's point distribution is much greater than that of the other player's, then the first player can be described as more (choose one most appropriate word) consistent / exciting.

3. Suppose a simple game involves rolling a fair, six-sided die that has one of the numbers 1, 2, 3, 4, 5, 6 on each side. The player picks a number from 1 to 6 before the die is rolled. The player wins \$8 if the number he picks is rolled, otherwise he loses \$2. Find the expected value and standard deviation of the game.

4. To make the game from the last problem more suspenseful, the same die will be rolled twice. If both times the number rolled is the same as the one the player picks, the player wins \$58, otherwise he loses \$2. Find the expected value and standard deviation of the game.

5. What is the moral of the last two problems combined?

7.6 HOW MUCH IS A LIFE WORTH?

The concept of expected values has applications beyond the ones that involve obvious numbers. We may not realize it, but a lot of our decisions are made based on the concept of expected values. The decision could be either "right" or "wrong" depending on the values we assign to the outcomes.

Example 7.6.1. *Tom Sawyer is contemplating jumping from the roof of one house to the roof of the adjacent house. Suppose the distance between the roofs of the two houses is 15 feet and Tom Sawyer never fails to clear 15 feet in practice, except when he has a cramp right before take-off, which rarely happens. And "rarely" here has been quantified by his nerdy friend Forrest Gump as "once in the past 300 attempts".*

So Tom figures he would fail with probability $1/300 \approx 0.003$, and succeed with probability $299/300 \approx 0.997$. Let F be the price of failure (injury, embarrassment, etc.) and S be the reward of success (confidence, respect from peers, a girl's heart, etc.), the expected value of his jump would then be $0.003F + 0.997S$. The jump is a "good" jump so long as the expected value is positive. Whether the expected value is positive or negative depends on the individual making the decision. One person (for, example, Tom himself) could say $F = -\$1,000,000$ and $S = \$10,000,000$, making the expected value $\$9,967,000$ and the jump totally worthwhile, or actually very attractive. Another person (for example, Tom's grandma) could assign a number like $-\$100,000,000$ to F and only $\$100$ to S, making the expected value $-\$299,900.30$ (that's NEGATIVE), and the jump a totally stupid act◄

It is no different from when a stuntman jumps over a row of cars on a motorcycle. The probability of success is estimated based on the laws of physics. Is a $300,000 prize worth the risk of say, falling off the bike or crashing to the ground with a 0.1% probability? If $300,000 doesn't look attractive, $3,000,000 may arouse some interest. But if the jump is from the Earth to the Moon, then no amount of money will lure anyone to try because the probability of success is zero—this leads us to our next observation: When a well-established man does something "stupid" that ruins his career, the public always asks "why would somebody throw all the success away for that?" The reason could be that the man assigned too high a value to the probability of getting away with the act, or too high a value to the rewards of the act, and so in his mind the "stupid" act was actually worth the risk. If it were certain that any "stupid" act would be exposed, then the number of well-established people who act stupidly would decrease dramatically.

∴∴ **EXERCISES 7.6** ∴∴

1. You are a skilled mountain / rock climber and there is this high mountain in your backyard on which some highly prized mushrooms grow. A trip to search for a mushroom (assume the mushrooms are loners and you are lucky if you can find one—finding two on the same trip is impossible) takes a week and the probability of finding and successfully bringing a mushroom back is 0.05. You can use the time and energy spent on each trip to do other work that provides a guaranteed income of $1000. There is also a slight chance that you

will get injured on a mushroom searching trip and never be able to work again, and "slight chance" here means a probability of 0.002. A mushroom commands a market value of $45000. Should you be out there looking for mushrooms? The answer of course, depends on how much value you place on your ability to work for the rest of your life. Assign some values to your ability to work for the rest of your life, do the computations, and share your observations.

2. I am standing in front of a single-plank bridge that goes from one side of a deep, roaring river to the other side. Some people have successfully crossed the river before, but those who fell were never seen again. The probability of successfully crossing the river is 0.01. The questions below have no definitive answers—you will have to make some assumptions in order to reach a conclusion.

   (a) Should I try to cross the river if a roaring lion is chasing me and the probability of me fighting off the lion is 0.001?

   (b) Should I try to cross the river if there is no danger behind me but there is $30 million worth of gold on the other side that will be mine if I can successfully cross the river?

3. Projects and class discussions assigned by the instructor or suggested by you.

7.7 THE DARKER SIDE OF MATHEMATICS

Math is a double-edged sword. It can be used for good, and also bad.

Example 7.7.1. *We have seen how the number 2 can be used to help people more efficiently find a contaminated barrel of whisky among 100 barrels (section 6.3). The same principle can be used to scam people.*

Suppose I have paid $1,000 for 100,000 email addresses from some unethical Internet merchant. My specialty is sports betting (well, you will see this is not necessarily true). Suppose it is basketball season now and tomorrow there is a game between two teams. I send an email to 50,000 email addresses predicting one outcome, and another email to the remaining 50,000 email addresses predicting the opposite outcome. After the game is played, 50,000 people will have been somewhat impressed by me because my prediction was correct. The other 50,000 people who received the *incorrect prediction are not on my mailing list any more. The next game is tomorrow, and I just repeat my scheme again: prediction of one outcome to 25,000 people, prediction of the opposite outcome to the other 25,000 people. After the game is played, my mailing list will have 25,000 people who have seen me make a correct prediction twice in a row.*

I will be patient and do it one more time so I can cultivate some loyal followers. After the third game is played I will have 12,500 people on my mailing list. These people have seen me make a correct prediction three times in a row. Some of them know basic probability and can find the probability of such a feat by someone who does it by pure guessing to be $(0.5)(0.5)(0.5) = 0.125$, which is not really, really low but kind of low so most likely this guy knows something, you know? They figure.

Before the next game, I send an email to the 12,500 people on my mailing list, asking for $100 from each person for the next prediction.

Suppose 10% of the people actually trust me, each of them gives me $100. I collect the $125,000 and close my business.

It is a made-up story, but you can see the principle behind it. And in the real world, people have been scammed by much simpler schemes. Don't be a victim◄

Example 7.7.2. *Here we look at the amazing power of exponential growth, and why the pyramid scheme is evil.*

Take a small amount of money, say $1, and double it every week. How long does it take to reach $1,000,000? Let's see, from $1 to $1,000,000... maybe a few years or decades at least? Well, it takes only 20 weeks, or less than five months. Here is how: Week 0: $1. Week 1: $2. Week 2: $4. Week 3: $8... We can see that the amount at week n is 2^n. So at week 19, the amount is $2^{19} = \$524,288$, more than half a million dollars. At week 20, the amount is $2^{20} = \$1,048,576$, more than one million dollars. The moral of the story: Never promise your children you will double their allowance every week if they behave. You promise to double their allowance every decade if you want to keep your retirement money.

What about pyramid schemes?

A pyramid scheme's basic structure goes like this: someone tries to recruit you to be part of a GREAT TEAM that sells some GREAT products (such as volcanic ash shampoo and poisonous lizard oil). Your job is easy. You need to maybe sell some of the products to your neighbors but most importantly, recruit some more people into the team. Every time your recruits make some money, you get a percentage of their profit. So the more people you can recruit, the more money you can make. To join the TEAM, you need to spend a couple thousand dollars to buy some of the GREAT products first... but don't worry, they are GREAT products and you will sell them in no time. Well, it is the super connected era and you have 1,000 virtual friends and maybe four or five real friends so you figure it shouldn't be too hard for you to convince maybe, 7 people, to join you.

But it is hard.

If you can get 7 people, and each of the these 7 people gets 7 brand new people, and each of these 7 third generation people gets 7 some other brand new people... where are we now? Language is confusing, let's do the math:

The team	Members in the team
You	1 person
Under you	7 people
Under under you	$7 \times 7 = 49$ people
Under under under you	$49 \times 7 = 343$ people (looking ridiculous already)
Under under under under you	$343 \times 7 = 2401$ people (theoretically still possible)
Under under under under under you	$2401 \times 7 = 16807$ people (if you is teacher ☉☉)
Under you six times	$16807 \times 7 = 117649$ people
Seven times under you	$117649 \times 7 = 823543$ people
Under eight times you	$823543 \times 7 = 5764801$ (you president?)
Nine	$5764801 \times 7 = 40353607$ (you rock star?)
Teen	$40353607 \times 7 = 282475249$
Eleveen	$282475249 \times 7 = 1977326743$
Twelveen	$1977326743 \times 7 = 13841287201$ (PEOPLE?)

If we add all the numbers above together, that's the size of your TEAM, which is greater than $16,000,000,000$ *(this is a geometric series so you can use formula (1.2.2) from chapter 1 to find the sum). The world's population was about* $7,600,000,000$ *as of the year 2017. So congratulations, you have assembled a sales army that includes all the people on planet earth and then some Martians and maybe some Venusians... but then why are you selling stuff? You might as well just rule the galaxy.*

Pyramid schemes are unsustainable. How many people in the world would be interested in volcanic ash shampoo and poisonous lizard oil? Even if everyone wanted some of your great products, how would you be able to produce them? The only people who benefit financially from such a scheme are the few people who start such a scheme, and they get their money not from selling great stuff, but from selling junk to people they recruit.

These types of schemes were more common in the last century, and are relatively rare today— maybe because people have gotten smarter (especially if people have read this book). But new schemes will appear and old schemes will reappear under different guises. Learning and practicing mathematics will give you a better chance at spotting an evil schemer when you do see one◄

: : : **EXERCISES 7.7** : : :

1. My nickname is FivePercent and I am a Ponzi schemer. Me and you are buddy buddies and you have under your mattress some spare moneys. I say to you "I have connections in high places that guarantee 5% of return every month, but the minimum investment is $1000." You do your calculus in your head and figure 5% per month amounts to a lot in a year and decide to give me $1000 to invest. One month later I give you $50 and you are impressed. So you tell two buddies of yours about me and they each give me $1000. At the end of the second month I give you and the two buddies of yours a total of $150. The two buddies of yours may introduce me to more people, but we all know this can't go on forever from the pyramid scheme example, so let's assume the number of my customers stops at three. At the end of the third month I give you three another $150 hoping you can get me more investors. At the end of the fourth month I give you three another $150 still hoping you can get me more investors. When no new investors join, I disappear.

    (a) How much money have you lost? How much money has each of your two buddies lost?

    (b) Assume I never invest any of the money because I know nothing about investment, and I actually spend $300 of the money on my personal necessities every month. How much money is in the "investment account" at the end of the first month, the second month, the third month, and the fourth month?

    (c) How much longer can I act like a hero in front of you guys if I don't disappear and assume no new customers are fooled by me?

2. Projects and class discussions assigned by the instructor or suggested by you.

7.8 ADDITIONAL PROBLEMS

In this section we present or analyze some easy but surprising, or not so easy but surprisingly solvable problems.

▶**PERCENT** and percent change are a simple concept, so simple they can cause people to make embarrassing mistakes—embarrassing in the "I can't believe I missed that" kind of way. The percent change formula is

$$\text{Percent Change} = \frac{\text{Newer number} - \text{Older number}}{\text{Older number}} \cdot 100\%$$

We know 100% = 1, so the 100% in the formula is there to remind us we want the answer to be written as a percent. For example, if Tim was 150 cm tall a year ago, and he is now 165 cm tall, then the percent change in Tim's height over the past year is $(165 - 150)/150 = 10\%$.

See if you can solve the following problems.

1. If a number p is 25% more than another number q, then q is _____ less than p.
   <u>Ans:</u> 20%. If you jumped at 25%, gotcha!

2. There are 99 dogs and one cat in a room. Obviously the dogs represent 99% of the total population. The cat surveys the room and feels there are way too many dogs, so he kicks a bunch of dogs out of the room. After that the dog population drops to 98%. How many animals are in the room?
   <u>Ans:</u> Population of dogs is down by 1% so maybe a few dogs left? Surprisingly, 50 dogs have to leave for this to happen. The room now has 50 animals: 49 dogs and 1 cat.

3. You paid the county $100 in property tax, but later were told that the county overcharged everyone by 25%. How much money should the county return to you?
   <u>Ans:</u> $25, right? No, $20 ◄

▶**AVERAGE** is another concept we need to be careful about. By "average" here we mean the most commonly used "arithmetic average". Suppose you go from point A to point B averaging 40 miles per hour. On the return trip (the exact same route reversed, so same distance) from point B to point A you average 60 miles per hour. What is your average speed for the round trip?
<u>Ans:</u> The answer is NOT 50 miles per hour. It is 48 miles per hour◄

▶**IN** the 2018 World Cup held in Russia, 32 teams entered the tournament. The 32 teams were divided into 8 groups of 4 teams each during the group stage. Within a group each team played every other team exactly once, and two teams from each group advanced to the next stage called the knockout stage. In the knockout stage, two teams played one game and the winner advanced to the next round. This process continued until there were 4 teams left. These 4 teams were divided into 2 groups and one game was played in each group. The winners of the two groups would go on to play one game that determined the champion and the second place winner, and the losers of the two groups would play one game to determine the third place winner. Answer the following questions:

(a) How many games were played during the tournament?

   - Within each group, $C(4, 2) = 6$ games were played. There were 8 groups, so the number of games played $= 6 \times 8 = 48$.

   - Now there were 16 teams left, 8 games played.

- Now 8 teams left and 4 games played.

- 4 teams left now and they are divided into two groups, 2 games were played.

- Two teams played for the 3rd and the 4th places, and two teams played for the 1st and 2nd places, 2 games played.

Ans: The total number of games played is the sum of all the games played we listed above: $48 + 8 + 4 + 2 + 2 = 64$.

(b) How many games did the championship team play?
Ans: $3 + 1 + 1 + 1 + 1 = 7$

(c) How many outcomes are possible for the entire tournament?
Ans: Each game has two possible outcomes, the total number of outcomes is $2^{64} = 18446744073709551616$ ◀

▶**ON** a remote planet the land is divided into three kingdoms X, Y, and Z, and the kingdoms are ruled by kings Xing, Ying, and Zing, respectively. Z is the most powerful, it can defeat either Y or X if it engages them one at a time. Y is the second most powerful and can defeat X if they get into a war without outside interference. X is the least powerful so it must consider its every move carefully in order to avoid annihilation. The three kings get into some heated arguments over some philosophical differences they post in the Clouds (real clouds in this case, not our mysterious "cloud computing" cloud in the year 2000-something on planet Earth). Now all-out wars are unavoidable. Z's plan is straightforward: it will "swallow Y in whole first, then consume X as a dessert" (bold statement from the most powerful king Zing). Y's plan is to persuade X to join forces with Y so they can fight off Z first, and see what happens after that. X, being the least powerful, has to weigh its decisions. X can adopt

- Strategy A: Let Z and Y fight their war and hope Z will be weakened to such a degree after the war that X can actually take Z out (a possibility the powerful Zing fails to recognize or refuses to acknowledge); or

- Strategy B: Join forces with Y, hopefully to defeat Z, but then face the possibility that Ying will find some excuse to start a war with X after their joint victory over Z.

Determine whether strategy A or strategy B will provide X with a better probability of surviving the wars if the following numbers are available to X:

$$Pr(\text{Z beats Y}) = 1$$
$$Pr(\text{Z weaker than X after defeating Y}) = 0.3$$
$$Pr(\text{X+Y defeats Z}) = 0.8$$
$$Pr(\text{Y finds excuse to start war with X after joint effort}) = 0.6.$$

Solution. Suppose X's goal is to survive the wars. The probability that X survives under each strategy is given below.

Strategy A: With probability 0.3 Z will be weaker than X after the war with Y, in that case X can defeat Z and survive. So the probability of survival equals 0.3.

Strategy B: For X to survive when adopting this strategy, X and Y must defeat Z first, which has a probability 0.8 of happening. After that Y must not start a new war with X, which has a probability 0.4 of happening. So the probability that X survives under this strategy equals $0.8 \times 0.4 = 0.32$.

When we compare the two probabilities above, we see that strategy B is a better strategy for X◄

▶**SOME** day in July 2019, I saw this little column in a newspaper. It seems that pyramid schemes are still going on and some people are still being victimized by, or victimizing, some other people :(

High-living parents secret

DEAR ABBY: JEANNE PHILLIPS

DEAR ABBY: My sister and I recently found out (through the internet) that my mother and stepfather have filed for Chapter 7 bankruptcy. A few months ago, Mom approached my sister (who's an attorney) asking about the effects of bankruptcy "for a friend."

My sister and I are now struggling with this information because my mom and stepdad promote a direct sales business where they advertise their multiple cars and lavish lifestyle as a result of the business profits. Should we let them know that we know about the bankruptcy and, if so, how should we handle this situation? — Struggling Sisters

DEAR STRUGGLING: You and your sister the attorney should go to your mother and stepdad and tell them the cat's out of the bag. They may need help extricating themselves from the company they have been promoting. Many people have been caught up in shady direct sales schemes and wound up with garages filled with product they couldn't sell. Whether your mother and stepdad are victims or perpetrators remains to be seen.

DEAR ABBY: My boyfriend and I have a joint membership at our local gym. Today the gym owner asked him if I was his mother. It upset me to the point of tears. We are not the same race. He is fit; I'm not, but we are both in our early 30s.

Why do people ask rude questions when a simple check of paperwork would satisfy their curiosity? I feel I should say something to her like, "Mind your own business." How do I get over this because I still would like to attend her gym? — Working Out in the Midwest

DEAR WORKING OUT: That gym owner could have lost two clients by asking that ill-advised question. Because you would like to continue patronizing the establishment, refrain from telling her to mind her own business.

◄

~Exercise Answers~

EXERCISES 1.1

1. 1.2   3. 24   5. 41   7. 3

2. 124.2   4. 0.4   6. 16920

8. (a) 1   (c) 1.5   (e) 2.25   (g) $5/6 \approx 0.83$
    (b) 2   (d) 1.25   (f) 2.5

9. (a) $7728   (b) $8592   (c) $9024   (d) $9628.80

10. (a) $1080

    (b) Seventeen payments of $171.11 each, plus one payment of $171.13—I know it's supposed to be EQUAL monthly payments, but in this case it is impossible to have equal monthly payments (why?).

11. 173005   12. 60%

EXERCISES 1.2

1. 4.5, 6.75   (c) More frequent compounding

2. (a) ≈ 145.68 5. $334.68

    (b) ≈ 2429.74 6. $\approx 8.4\%$ assuming interest is compounded annually

3. (a) 128

    (b) ≈ 170.67 7. Make sure you got it

4. (a) $\approx \$551.25$ 8. $\approx 9.4\%$

    (b) $\approx \$552.47$ 9. (D)

EXERCISES 1.3

1. 15 years

2. (a) Because the growth of the investment with the higher initial value follows a straight line, but the growth of the investment with the lower initial value follows an exponential curve

(b) After 3 years, mine = $5750, yours ≈ $4630.50, mine > yours. After 30 years, mine = $12500, yours ≈ $17287.77, mine < yours. So the answer lies somewhere between 3 and 30. We can try 16 years next and after that the time interval is further reduced. Keep going for a few more times and we will find the answer: eighteen years

3. 24 years

4. Enjoy

5. ≈ $20097.66

6. ≈ $2076.26 after 20 years; ≈ $546.21 after 30 years

EXERCISES 1.4

1. ≈ $36018.32

2. ≈ $37459.05

3. ≈ $20157.69

4. (a) ≈ $6333.79   (b) ≈ $260054.80

5. (a) ≈ $3070.59   (b) ≈ $236941.05

6. (a) 2688000
    (b) 2688000
    (c) 2688000
    (d) 2688000
    (e) 2688000
    (f) 2688000
    (g) 2744000
    (h) The results above show that a sequence of three 40%'s beats a sequence of 20%, 40%, and 60% in any order. Now verify this: A sequence of 30%, 40%, and 50% also beats a sequence of 20%, 40%, and 60%. (I know, this is the answers section so you expect to just get answers. But learning occurs everywhere and you are rewarded for coming here and reading this little part. People who skip it don't know as much as you do now.)

EXERCISES 2.1

1. distinct, elements, members

2. Yes

3. {7, 14, 21, 28, 35, 42, 49}

4. element

5. (a) No
    (b) Yes
    (c) No
    (d) No. Ø or { }, not {Ø}

EXERCISES 2.2 & 2.3

1. (a) U (c) B (e) $\{1,2,3,5,7,9\}$
   (b) $\emptyset$ (d) $\{3,5,7\}$ (f) C

2. (a) U (d) A (g) $\{\bigstar, \spadesuit, \diamondsuit\}$
   (b) $\emptyset$ (e) $\{\bigstar, \spadesuit, \clubsuit, \diamondsuit\}$
   (c) A (f) $\{\bigstar, \spadesuit, \clubsuit, \diamondsuit\}$

3. (a) {Ali, Bre}
   (b) {Irene, John}
   (c) {Ali, Bre, Irene, John}
   (d) Same as above
   (e) OR = union
   (f) {Ali, Bre, Cory, David, Elvis}
   (g) {David, Elvis, Frank, Gina, Helen, Irene, John}
   (h) {David, Elvis}
   (i) Same as above
   (j) AND = intersection
   (k) {Ali, Bre, Gina, Helen, Irene}
   (l) {Ali, Bre, Irene}

4. This is the most smartest observation I have never made

5. (A) (iii)   (C) (v)   (E) (ii)   (G) (v)   (I) (ix)   (K) (viii)
   (B) (iv)   (D) (vi)   (F) (i)   (H) (vi)   (J) (vii)   (L) (viii)

EXERCISES 2.4

1. (a) T   (c) T   (e) T   (g) F   (i) T   (k) T
   (b) T   (d) T   (f) T   (h) F   (j) T

2. Hint: Replace A with A' and B with B' in the given formula.

3.

4.

5. Construct your own argument, then compare with the argument given in the book.

EXERCISES 2.5

1. OR=UNION; AND=INTERSECTION

2. (a) {red Chevrolet Corvette, red Chevrolet Camaro, blue Chevrolet Cavalier}

   (b) {red Chevrolet Corvette, red Chevrolet Camaro, red Ford F-150}

   (c) {red Chevrolet Corvette, red Chevrolet Camaro, blue Chevrolet Cavalier, red Ford F-150}

   (d) They are the same

   (e) {red Chevrolet Corvette, red Chevrolet Camaro}

   (f) Same as above

   (g) 3, 3, 4, 2

   (h) $4 = 3 + 3 - 2$

3. 260

4. (a) 29   (b) 24   (c) 26   (d) 15

5. 7

6. (a)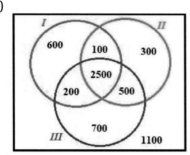

   (b) 1100

   (c) 1000

   (d) 700

   (e) 4900

   (f) 1600

   (g) 3300

   (h) 800

7. (a)   (b) 21   (c) 4   (d) 17

EXERCISES 3.1

1. Really?

2. (a) $(-1, -1, 2)$ (b) $(-1, 3, 8)$ (c) $(3, 7)$ (d) $(1, 0, 2, 9)$

3. (a) $(18/25, 11/25, -32/25)$ (c) $(31/4, -21/4)$

   (b) $(-103/75, 173/75, -16/75)$

4. (a) $(17,0), (-1,-6), (20,1)$ (c) $(1,2,3), (-1,-4,13), (0,-1,8)$

   (b) $(7,19,0), (6,20,1), (8,18,-1)$

5. 20 gallons of 40% and 30 gallons of 90%   7. 240 \$10's, 120 \$6's, 20 \$0's

6. \$$40M$, \$$30M$, \$$20M$ 8. (A)   9. (C)

EXERCISES 3.2

1. objective function; minimize   3. mathematical optimization

2. constraints

EXERCISES 3.3

1. (A) (vii)   (C) (i)   (E) (ii)   (G) ((vi)

   (B) (iii)   (D) (v)   (F) (iv)   (H) (viii)

2. (a)   (b)   (c)   (d)

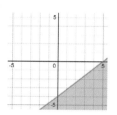 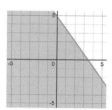 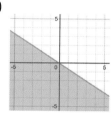 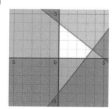

3. (a) Vertices of feasible region: $(0,0), (1,0), (3,2), (2,4), (0,6)$. Max Obj $= 27$ at $(3,2)$

   (b) 34 at $(2,4)$

   (c) 2 at $(1,0)$

   (d) -18 at $(0,6)$

   (e) Vertices of feasible region: $(1,0), (3,0), (3/2,3), (0,4), (0,2)$. Max Obj $= 15$ at $(3,0)$

   (f) 16.5 at $(3/2,3)$

   (g) 9 at $(0,4)$

4. (a) Let $x =$ time spent shadow boxing, $y =$ time spent rope jumping, $z =$ time spent icy water bathing

   (b) Max $300x + 500y + 50z$

   (c) Subject to $\begin{cases} x + y + z \le 60 \\ z \le 10 \\ x - y + z \ge 0 \\ x, y, z \ge 0 \end{cases}$

5. (a) Let x, y, z be the numbers of dishes A, B, C, respectively

(b) Max $5x + 7y + 12z$

(c) Subject to $\begin{cases} 2x + 4y + 5z \leq 110 \\ 3x + 5z \leq 80 \\ x + 2y + 3z \leq 60 \\ x, y, z \geq 0 \end{cases}$

6. (a) Let v, w, x, y, z be the amounts of foods V, W, X, Y, Z needed, respectively

(b) Minimize $C_v v + C_w w + C_x x + C_y y + C_z z$

(c) Subject to $\begin{cases} a_{11} v + a_{12} w + a_{13} x + a_{14} y + a_{15} z \geq 500 \\ a_{21} v + a_{22} w + a_{23} x + a_{24} y + a_{25} z \geq 400 \\ a_{31} v + a_{32} w + a_{33} x + a_{34} y + a_{35} z \geq 300 \\ a_{41} v + a_{42} w + a_{43} x + a_{44} y + a_{45} z \geq 200 \\ v, w, x, y, z \geq 0 \end{cases}$

7. • Let x, y, z be the amounts of good stuff from A, B, C, respectively

• Minimize $5x + 4y + 3z$

• Subject to $\begin{cases} x + y + z = 800 \\ x, y, z \geq 100 \\ x, y, z \leq 400 \\ 0.47x + 0.55y + 0.80z \geq 0.50 \times 800 \\ 0.47x + 0.55y + 0.80z \leq 0.60 \times 800 \end{cases}$

8. • Let x be the amount of timber and y the amount of stone ordered from A. It follows that the amount of timber from B is $500 - x$ and the amount of stone from B is $900 - y$

• Minimize $5x + 6(500 - x) + 4y + 2(900 - y)$, or equivalently, $4800 - x + 2y$

• Subject to $\begin{cases} x + y \geq 200 \\ x + y \leq 1150 \\ x \leq 500 \\ y \leq 900 \\ x, y \geq 0 \end{cases}$

9. No

EXERCISES 3.4

1. (a) 2×3 (b) 3×2 (c) 1×4 (d) 3×1 (e) 1×1

2. (a) -5 (b) 2 (c) 2nd, 3rd

3. 1, 4, 5, 2, 3, 6.

4. (a) $\begin{bmatrix} 2 & -5 \\ 3 & -2 \end{bmatrix}$

    (b) $\begin{bmatrix} 0 & -2 \\ 12 & -14 \end{bmatrix}$

    (c) Undefined. Can't add a number to a matrix

    (d) Undefined. Matrices are of different sizes

    (e) $\begin{bmatrix} 7 \\ -12 \\ 8 \end{bmatrix}$

    (f) $\begin{bmatrix} 8 & -13 \\ -30 & 41 \end{bmatrix}$

    (g) $\begin{bmatrix} 27 & 2 \\ -50 & 2 \\ -11 & 18 \end{bmatrix}$

    (h) $\begin{bmatrix} 8 & 10 & 12 \\ -4 & -5 & -6 \\ 0 & 0 & 0 \\ 48 & 57 & 66 \\ 60 & 72 & 84 \end{bmatrix}$

    (i) Undefined. Sizes don't match

    (j) $\begin{bmatrix} -2x+3y+5z \\ 11x+9y+13z \\ 7x-8y+16z \end{bmatrix}$

    (k) $\begin{bmatrix} x \\ y \\ z \end{bmatrix}$

    (l) $\begin{bmatrix} z \\ y \\ x \end{bmatrix}$

    (m) $\begin{bmatrix} x \\ z \\ y \end{bmatrix}$

    (n) Yes?

    (o) O, I see

    (p) $\begin{bmatrix} 44 \end{bmatrix}$

    (q) $\begin{bmatrix} 2x+4y+6z \end{bmatrix}$

    (r) $\begin{bmatrix} 2 & 6 & 10 \\ 4 & 12 & 20 \\ 6 & 18 & 30 \end{bmatrix}$

    (s) $\begin{bmatrix} 2x & 2y & 2z \\ 4x & 4y & 4z \\ 6x & 6y & 6z \end{bmatrix}$

    (t) Undefined

    (u) Undefined

    (v) $\begin{bmatrix} 30 & 10 \\ -20 & 30 \\ 40 & 30 \end{bmatrix}$

    (w) $\begin{bmatrix} -2x+3y+5z-1 \\ 11x+9y+13z-2 \\ 7x-8y+16z-3 \end{bmatrix}$

5. (a) $\begin{bmatrix} 14 & 15 & 13 \end{bmatrix}$. Same as adding the numbers in each column

    (b) $\begin{bmatrix} 7 & 7.5 & 6.5 \end{bmatrix}$. Same as taking the average of numbers in each column

    (c) $\begin{bmatrix} 23 \\ 19 \end{bmatrix}$. Same as taking the sum of the numbers in each row

    (d) $\begin{bmatrix} 23/3 \\ 19/3 \end{bmatrix}$. Same as taking the average of the numbers in each row

6. By company (A). By month (C)

7. (C)

8. (a) $\begin{bmatrix} 9 & 7 & 5 \\ 8 & 6 & 4 \\ 3 & 2 & 1 \end{bmatrix}$

    (b) $\begin{bmatrix} 5 \\ 1 \\ 2 \end{bmatrix}$

    (c) Undefined

    (d) $\begin{bmatrix} a & b & c \\ x & y & z \end{bmatrix}$

9. (a) T

    (b) T

    (c) F

10. (a) $(2, -1, -1)$

(b) $(6, -1, -3)$

(c) $(16, 15, 21)$

(d) $0.28, -1.89, -0.07)$

(e) $(-15, 8, 13)$

(f) $(1, 0, 2)$

(g) $(a + 2b + c, 2a - b + 3c, 2a + 2b + c)$

11. True

EXERCISES 4.1

1. 12

2. 1814400

3. 676

4. 10000

5. 9000

6. 1572120576

7. 16

8. 16

9. 16

10. You are right

11. (a) 16

(b) 81

EXERCISES 4.2

1. (a) 720

(b) 120

(c) 24

(d) 6

(e) 2

(f) 1

(g) 1

(h) 720

(i) 720

(j) 360

(k) 120

(l) 30

(m) 6

(n) 1

(o) 1

(p) 6

(q) 15

(r) 20

2. Both give the number of ways to divide 1000000 objects into two groups, one with 999998 objects in it and the other with 2 objects in it

3. 120

4. 720

5. 66

6. 1320

7. 126

8. $C(4, 0) + C(4, 1) + C(4, 2) + C(4, 3) + C(4, 4) = 16$

EXERCISES 4.3

1. 1684800

2. (a) $12! = 479001600$

(b) $10!3! = 21772800$

(c) $9!4! = 8709120$

(d) $6!3!5! = 518400$

(e) $3!3!4!5! = 103680$

(f) $3!4!5! = 17280$

3. 15

4. 64

5. 350

6. $C(18, 5)C(13, 6)C(7, 7) \cdot 5 \cdot 6 \cdot 7 = 3087564480$

7. $1 \cdot C(17, 5) \cdot 5 = 30940$

8. 48

9. (a) 11

(b) 66

(c) 72

EXERCISES 4.4

1. (a) $\dfrac{10!}{3!\underline{7!}}$ (c) $\dfrac{P(10,3)}{\underline{3!}}$ 4. 14702688

    2. 2522520   5. 14702688

    (b) $\dfrac{10!}{1!1!1!\underline{7!}}$ 3. 36756720   6. 360360

EXERCISES 5.1

1. (a) $\{1,2,3,4,5,6\}$ (b) $\{2,4,6\}$ (c) $\{3,4,5,6\}$

2. (a) $\{HHH, HHT, HTH, THH, TTH, THT, HTT, TTT\}$

    (b) $\{HHH, HHT, HTH, THH\}$

    (c) $\{HHH, HHT, THH, THT\}$

3. 1   5. 0   7. No   9. 0.10

4. 1   6. No   8. No   10. 0.17

EXERCISES 5.2

1. 5/11   2. 5/11   3. 1/4   4. 5/14

5. (a) 1/84   (e) 64/84

    (b) 18/84   (f) They form the sample space

    (c) 45/84   (g) Use formula (5.2.2)

    (d) 20/84   (h) Easier to work with

6. (a) 64   (b) 15/64   (c) 63/64

7. (a) 64   (b) 15/64   (c) 63/64   (d) Same, why?

8. (a) 64   (b) 15/64   (c) 63/64   (d) Thx 4 lecture

9. $(7!/(1!2!4!))/3^7 = 35/729$

10. (a) 17/47   (b) 13/47   (c) 5/47   (d) 25/47   (e) 22/47

11. 21%

EXERCISES 5.3

1. $\Pr(E|F)$

2. (B)

3. (a) $\Pr(D|P)$ (c) $\Pr(R \cap P)$ (e) $\Pr(R \cup D)$

    (b) $\Pr(P|R)$ (d) $\Pr(R \cap D)$ (f) $\Pr(D|R)$

4. (a) 1/2   (b) 1/3

5. (a) 0.3   (e) 2/3

    (b) 3/5   (f) 1/3

    (c) 3/7   (g) Yes

    (d) 4/7   (h) No, because $\Pr(E)|\Pr(F) \neq \Pr(E \cap F)$

6. (a) 0.42   (c) 0.4

    (b) 0.3   (d) Yes, because $\Pr(F)|\Pr(C) = \Pr(F \cap C)$

7. (a) 0.30   (c) 6/7

    (b) 3/4   (d) No, because $\Pr(D)|\Pr(G) \neq \Pr(D \cap G)$

8. Yes, Dr. Stubborn should change his mind. The two events are not independent

9. (a) 0.995   (b) 0.994   (c) 0.997   (d) MORE   (e) reflecting

10. (a) 0.006   (b) 0.092   (c) 0.398   (d) 0.504   (e) 0.994

EXERCISES 5.4

1.

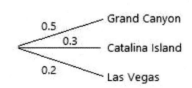

2. (a) ↘   (b) 0.375   (c) 0.05

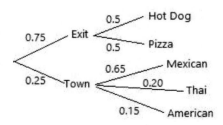

3. (a) ↘ (b) 0.2 (c) 0.24 (d) 0.336 (e) 0.664

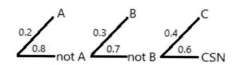

4. (a) ↘ (b) 1/7 (c) 1/7 (d) 1/7 (e) 1/7

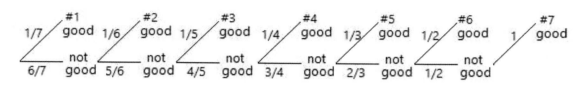

5. Mathematically equivalent to the previous problem

6. 7/13 (b) 33/47 11. (a) 78%

7. 9/23 9. 28/99 (b) ≈ $63.64

8. (a) 2/11 10. 117/171, 40/171, 14/171 (c) Ask if you don't see it

EXERCISES 6.1

1. 0.6 3. random variable

2. 0.1 4. $\Pr(R) = \Pr(S) = \Pr(P) = 1/3$

5. (a)

X	Probability
0	35/1365
1	280/1365
2	588/1365
3	392/1365
4	70/1365

(b)

Y	Probability
0	70/1365
1	392/1365
2	588/1365
3	280/1365
4	35/1365

(c) Since $X + Y = 4$,
$\Pr(Y = 0) = \Pr(X = 4)$
$\Pr(Y = 1) = \Pr(X = 3)$
$\Pr(Y = 2) = \Pr(X = 2)$
$\Pr(Y = 3) = \Pr(X = 1)$
$\Pr(Y = 4) = \Pr(X = 0)$

6.

X	Probability
0	0.006
1	0.092
2	0.398
3	0.504

7. two, success, failure

8. successes, p, n

9. (a) ≈ 0.282

(b) ≈ 0.526

(c) ≈ 0.944

10. (a) ≈ 0.9998530974

(b) $0.1^{10} = 0.0000000001$

11. (a) $C(10,2)(1/6)^2(5/6)^8 \approx 0.291$

(b) ≈ 0.515

EXERCISES 6.2

1. long term

2. No, we need to know how many credits are A's and how many are B's

3. 4.8

4. 0.6

5. $0.10, −$0.10

6. $9.36

7. 47.5%

8. 47.5. It is easy but you want to see the math is the same as the one above

9. 43.5 minutes

10. False

11. $1340

EXERCISES 6.3

1. $3.60

2. 93%

3.   (a) Binomial distribution with $n = 10$ and $p = 1/4$.

    (b) 2.5

4.   (a) Binomial distribution with $n = 3$ and $p = 0.1$.

    (b) −10 Unicoins

5. −32.225 Unicoins

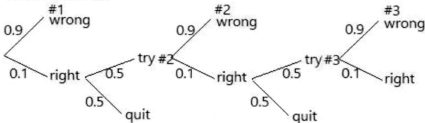

6.   (a)

X	Probability
0	7/99
1	35/99
2	42/99
3	14/99
4	1/99

    (b) 5/3.

7. Same as last one

EXERCISES 6.4

1. central tendency, spread (or dispersion)

2. increase

3. $\sigma^2 = 1.25$, $\sigma \approx 1.12$

4. $\sigma^2 = 0.32$, $\sigma \approx 0.57$

5. $\sigma^2 \approx 0.71$, $\sigma \approx 0.84$

6. $\mu = 50$, $\sigma^2 = 25$, $\sigma = 5$

7. $\mu = 30$, $\sigma^2 = 25$, $\sigma = 5$

8. $\mu = 15$, $\sigma^2 = 14.85$, $\sigma \approx 3.85$

EXERCISES 6.5

1. (a) significant

   (b) more

2. (a) Mean ≈ 5.11, median $= 5$

   (b) Mean $= 16.6$, median $= 5$

   (c) The mean can be influenced by a few extreme members in the population

3. A, B, C

4. A, C

5. (a) Both $= 5$

   (b) Group A, because the 5's comprises 70% of the population in A, but only 20% in B

6. median $= 106.5$, min $= 100$, max $= 110$, 1st quartile $= 102$, 3rd quartile $= 109$, range $= 10$, inter-quartile range $= 7$

EXERCISES 7.1

1. A $= 16/17 \approx 94\%$, B $= 1/17 \approx 6\%$. This is an over simplified explanation for why some businesses that don't seem too ethical always exist and may even survive for a long time. And by the way the initial distribution of 50%-50% doesn't affect the long term distribution.

2. $5/6 \approx 83\%$

3. $x = 23/73 \approx 32\%$, $y = 22/73 \approx 30\%$, $z = 28/73 \approx 38\%$

EXERCISES 7.2

1. 56

2. (a) 210   (b) 90   (c) 18   (d) 3/7   (e) 3/35

3. 30

4. From B to Z: 1 way; from Z to A: 21 ways. 1 × 21 = 21
   From B to Y without going through Z: 1 way; from Y to A: 15 ways. 1 × 15 = 15
   From B to X without going through Y or Z: 1 way; from X to A: 10 ways. 1 × 10 = 10
   Total from B to A: 21 + 15 + 10 = 46 routes
   Try using the same principle but go from A to B. It feels amazing when you get the same answer even though that's to be expected
   Alternatively, we can put the missing blocks back (see figure below). The total number of routes from B to A would be 56. Now take away all the routes from B to V to A (taking away V takes away all the routes from B to A that pass through the dotted lines), there are 10 of them. 56–10 = 46

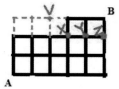

5. 105

EXERCISES 7.3

1. (a) OR

    (b)

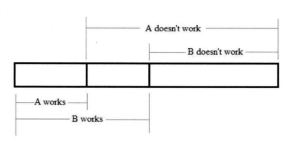

3.

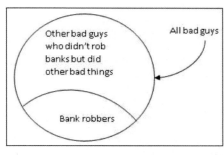

2.

4.

189

EXERCISES 7.4

1. (a) $M =$
$$\begin{bmatrix}
0 & 0 & 0 & 0 & 0 & 0 & 0 & 0 & 0 & 0 & 0 & 0 \\
0 & 1 & 1 & 0 & 0 & 0 & 0 & 0 & 0 & 0 & 1 & 1 \\
0 & 1 & 1 & 1 & 1 & 1 & 1 & 1 & 1 & 1 & 1 & 1 \\
0 & 0 & 1 & 0 & 0 & 0 & 1 & 0 & 0 & 0 & 1 & 0 \\
0 & 0 & 1 & 0 & 1 & 0 & 1 & 0 & 1 & 0 & 1 & 0 \\
0 & 0 & 1 & 0 & 0 & 0 & 1 & 0 & 0 & 0 & 1 & 0 \\
0 & 0 & 1 & 1 & 1 & 1 & 1 & 1 & 1 & 1 & 1 & 0 \\
0 & 0 & 1 & 1 & 1 & 1 & 1 & 1 & 1 & 1 & 1 & 0 \\
0 & 0 & 1 & 0 & 0 & 1 & 0 & 0 & 0 & 1 & 1 & 0 \\
0 & 0 & 1 & 0 & 0 & 0 & 0 & 1 & 0 & 1 & 1 & 0 \\
0 & 0 & 1 & 1 & 1 & 1 & 1 & 1 & 1 & 1 & 1 & 0 \\
0 & 0 & 1 & 1 & 1 & 1 & 1 & 1 & 1 & 1 & 1 & 0
\end{bmatrix}$$

(b) $AM =$
$$\begin{bmatrix}
0 & 0 & 1 & 0 & 0 & 0 & 1 & 0 & 0 & 0 & 1 & 0 \\
0 & 0 & 0 & 0 & 0 & 0 & 0 & 0 & 0 & 0 & 0 & 0 \\
0 & 1 & 1 & 1 & 1 & 1 & 1 & 1 & 1 & 1 & 1 & 1 \\
0 & 0 & 1 & 1 & 1 & 1 & 1 & 1 & 1 & 1 & 1 & 0 \\
0 & 0 & 1 & 1 & 1 & 1 & 1 & 1 & 1 & 1 & 1 & 0 \\
0 & 1 & 1 & 0 & 0 & 0 & 0 & 0 & 0 & 0 & 1 & 1 \\
0 & 0 & 1 & 0 & 0 & 0 & 1 & 0 & 0 & 0 & 1 & 0 \\
0 & 0 & 1 & 0 & 0 & 1 & 0 & 0 & 0 & 1 & 1 & 0 \\
0 & 0 & 1 & 1 & 1 & 1 & 1 & 1 & 1 & 1 & 1 & 0 \\
0 & 0 & 1 & 1 & 1 & 1 & 1 & 1 & 1 & 1 & 1 & 0 \\
0 & 0 & 1 & 0 & 0 & 0 & 0 & 1 & 0 & 1 & 1 & 0 \\
0 & 0 & 1 & 0 & 1 & 0 & 1 & 0 & 1 & 0 & 1 & 0
\end{bmatrix}$$

(c)

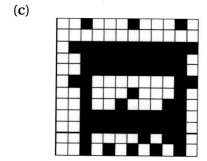

(d) A long and tedious problem like this will leave a longer lasting impression

EXERCISES 7.5

1. boring   2. exciting   3. −0.33, 3.73   4. −0.33, 9.86

5. When the expected value is held constant, a larger winning amount and a larger variance

are related (the player wins a higher prize when he does win, but wins less often)

EXERCISES 7.6

1. It is even in terms of monetary rewards between doing regular work and searching for mushrooms if you place your ability to work at a value of $625,000 per week. In real life there are other variables to consider, for example, how much value you place on the anxiety your family has to bear every time you go looking for mushrooms

2. (a) If I just want to stay alive, then I should try to cross the river because crossing the river gives me a higher probability of staying alive. If I am really angry at the lion and think I must protect my territory at all cost, I should try to karate chop the lion to death

   (b) Given that the probability of successfully crossing the bridge is 0.01, the expected reward is only $0.01 \times \$30,000,000 = \$300,000$, and that's if I decide to stay on the other side of the river forever. If I think my life on this side of the river is worth more than that, then I shouldn't cross. If I want to make it back to where I started, then the expected reward is only $0.01 \times 0.01 \times \$30,000,000 = \$3,000$ (that's not much if you ask me). To be fair, some people did cross some unknown bodies of water in history, though, and rediscovered something really great

3.

EXERCISES 7.7

1. (a) You lost $800, each buddy lost $850.

   (b) First month $650; second month $2,200; third month $1,750; fourth month $1,300

   (c) Less than three months

2.

~Epilogue~

What's written below are some random thoughts and random recollections.

Writing a book may seem like a lonely journey, but no one is completely alone. Everyone is connected to many people and many things directly, and even more people and more things indirectly.

I was very fortunate to have great parents and siblings when I was born. They have always been there to care for and look after me.

Then I grew up, got restless, and found a beautiful girl to marry. That's a great adventure and caused some heartache for many handsome young men—some of them are still handsome, and bitter, too.

Then we expanded our family and it was over a decade of uncertainty and anxiety, but also hope and joy. We were poor, but also full of jubilance!

I have so many stories to tell and so many people (some of them animals) to thank. Those stories will be told in separate volumes, as masterpieces, of course.

I will now proceed to list the people and events that I remember fondly. Some people may not remember me or even know me, some people have passed, some people will never know this little book exists and will never read this little piece. In a way, writing is selfish. I am writing things down to make myself feel better.

To my father, 黃碧沛, who passed away on Valentine's Day in 1982. I had so many things I wanted to tell him but never had the opportunity to. To this day, I still feel his love. He left behind many beautiful poems, and deep and loving memories in many people's hearts.

付与鼓浪屿的深情
碧沛
我很想把你连海抱走
轻轻地放在我的窗前
四季飘散着花的芬芳
夜夜灯山映红海面

我很想把你连海抱走
时刻带在我的身边
绿茵中升起缕缕琴声
曲桥上细诉绵绵情言

To my mother, 鄭英如, a strong and brave woman who has devoted her life to her family, children, and grandchildren. She found Jesus Christ late in her life and that has been a true blessing. But to hear her say it, the seeds were planted by an American pastor's wife way back when she was a skinny tiny ugly little girl in a small poor southern Chinese village. The American pastor even had a Chinese surname: 蔡, and better yet, both she and her husband spoke the local Chinese dialect, which is very different from the well-known Chinese dialects of Mandarin and Cantonese—that should send chills down your spine— it takes so much devotion to go to a foreign country, suffer the harsh living conditions, learn a nth foreign language, and give love and hope to the people who are deprived of them.

To my sisters, 黃以茜 and 黃彩笛. Even though my parents favored me when we were growing up simply because I was a boy, they were never jealous and never complained. They have loved and cared for me like a parent would a child.

To my wife, the only love of my life, my childhood idol, my elementary-school crush, the flying little bird. She ran so fast I couldn't catch her, so I ran in the opposite direction of the track and that's how I caught her. We have journeyed the world for decades and she has given me and taught me so much. Thanks honey.

To my son, such a joyful boy. I wish I wasn't as physically exhausted as I was so I could have spent more time with him when he was little. Not that we didn't spend time together, we did aplenty. I just wished it could have been more. Lessons learned here: don't hang out with big cats if you are a mid-sized cat. Translation: don't get into a Ph.D. program in mathematics if you are not mathematically talented. Anyway, now that I am old, we mostly do a little boring stuff called Tai Chi together.

To my grandfather and grandmother. I will always remember the love and caring they gave me, even though they were not expecting anything in return from a confused young boy. They were always resilient and self-reliant. I am very proud to be their grandson. Some love you received you can never repay because when you were receiving it you were too young or too stupid to realize what was happening.

To the several cats who had shared their lives with me. A true teacher leads by example, not by words. These cats did just that. Thank you for all the great lessons, not to mention the joys and tears you brought me. Kiss kiss whiskers. Meow meow.

On 17 February, 1982, on a bus from Xiamen to Hong Kong, I bumped into a young man named 滕志平. I was on my way to attend my father's funeral in Hong Kong, three days after his passing. Mr. Teng, maybe ten years older than me, kept me company, gave me a few lectures on Hong Kong and capitalism, and shared one pack of cigarettes with me through the night. It was only half a day in my life, but the memory of him has lasted through the years.

And just like Mr. Teng, many random people have done many kind things for me. I am forever grateful for these people. They made me kinder and more hopeful.

Thanks also to all the people who treated me unfairly or unkindly. They have certainly made me angrier, stronger, smarter, and funnier. Wish you guys the best all the same. Next time pick on somebody your own size, though. There is no glory beating up a small guy, you losers.

A few professors at the University of Arizona have changed the way I do mathematics. Hermann Flaschka showed me one can be math-smart and cool at the same time. Bruce Bayly showed me math can be done in a relaxed yet exciting manner. Oma Hamara gave me the confidence, trust, and encouragement that made me a competent teacher. Sidney Yakowitz generously gave me his time and energy in guiding me to become a more mature graduate student, and he cared greatly about my future, too. My dissertation advisor Moshe Shaked was a giant in the field of stochastic orders and reliability theory. He took me, a person with an average math IQ, under his wing and quietly gave me the guidance, with great patience and close attention, that led to the successful completion of my dissertation. He gave me some money, too, in the form of research assistantship, and I used part of that to buy, among other important things, some cheap beer. He didn't say much, but I could feel his heart. "Those who know don't talk. Those who talk don't know." Last but not least, it was awe-inspiring to be in two classes taught by Lai-Sang Young (Real Analysis I and Probability Theory). Who says Chinese women can't do mathematics?

Some students who took my classes at the University of Arizona, despite their young age at that time, showed genuine care about my teaching. Their friendliness, suggestions, and encouragement meant so much to me. How can one not like young people? They are the future and we should do our best to help them learn. They want to offer love and help, too.

Many students from Hong Kong and China who attended the University of Arizona when I was a graduate student there helped me in many, many ways. One can never make it without friends. Much, much gratitude!

To the loyal friends from my childhood in Xiamen, thanks for being loyal. Remember you are finally as old as I am now, so please try not to drink too much.

To my wonderful College of Southern Nevada (Community College of Southern Nevada prior to July 1, 2007) colleagues. Thanks for tolerating me and not talking behind my back. I have never done much for you guys but you can be sure of one thing: I love you all, especially the ones who drink, and the ones who are shorter than me. Special respect to Stan VerNooy, a very special colleague. The world is not as exciting without him. We had a special connection right from the beginning: I came to CCSN in 1996 driving a 1988 Dodge Lancer, he came to CCSN in 1997 driving a 1987 Dodge Lancer, like we were twin Lancer brothers Don Quixote and Don Quixote. We had some fierce email sparring over the years, those were equally special. Rest in peace, my friend!

Through the years I have learned a lot from my students, some at the University of Arizona and some at the (Community) College of Southern Nevada. Some super smart, some very funny, some sensitive, some highly ambitious, some seemed lost, and some actually cared about me. Thank you all.

And thank you especially to Trinity as she represents everything that is good. She also appreciates how the identity

$$\sum_{i=1}^{n} i^3 = \left(\sum_{i=1}^{n} i \right)^2$$

can be demonstrated by wooden blocks

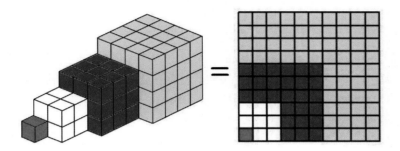

I still remember my first grade teacher Ms. Liu, although short haircut is all I remember about her appearance. I could never do wrong in her eyes. She was my teacher for only one semester, and that's probably good for me because she would have spoiled me for sure. Also special was my high school math teacher Mr. Liu Yu Liang. He emphasized understanding over mindless drilling—he could have easily been the founder of the reform math movement that went berserk in the 1990's in the United States. On the other end of the spectrum was Mr. Huang Han Qiang, another high school math teacher of mine. Rigid as robot but the discipline he instilled in me benefited me for the rest of my life.

To a dear friend and classmate from Xiamen University, whose life was forever changed one day because he tried to help his friend, me, when he delivered some documents to the university to keep me enrolled when I was in Hong Kong attending my father's funeral, but as a consequence missed some precious minutes in his graduate school entrance exam. Love you with all my heart. Believe me, it's better that you became an engineer rather than a scientist as you originally planned. As the saying goes, "those who can, engineer; those who can't, scientist!"

To Bruce Lee, for inspiring billions of people and changing history with his spectacular display of authentic artistry and creativity, unmatchable energy and power, and undeniable enthusiasm and charm.

To Albert Einstein, who taught me that if you can clean your face and body with the same soap, then there is no need to use two soaps.

I ate an orange the other day, it was good. A lot of people and work were involved in creating the orange and making it available for my enjoyment. When people say they are self-made, without any help from anyone, they are just joking, OK? I am sure most people don't have the ability to make a shirt from, say, cotton. And a few who do probably don't know how to build a car from scratch which they sometimes need in order to go places. And if some people can actually make a shirt from cotton and build a car from dirt, I am pretty sure they can't also dig up oil from the earth and turn it into gasoline for the car.

Thanks to Henry Wong for his meticulous poofreading. Any miss$pelling or errors in grammaRs in book now his risponsibilities.

$$> < \boxed{\mathfrak{ADDENDUM}} > <$$

Microsoft Word is not good for writing mathematics. The formulas in MS Word look ugly, and lately (summer 2019 is the current time), everything that I created with MathType has been subjected to random resizing in my Word documents. "Random" because I couldn't figure out if there is a formula Microsoft uses to mess up all my equations and formulas. I almost wanted to present this book "as is" in Word because the math equations are so bad it could be an instant classic.

And cross-referencing in Word is difficult, too. Also, there is always this uneasy feeling that your document may not look the same when a new version of Word comes out. That's the worst! You buy the newest version of Word, and when you open a once-beautiful document that you spent months or years to create, so many things look different or out of place.

That said, I did use MS Word and Paint to create some graphs and drawings. Among them, the opossum, the cat and the dog, the Bruce Lee beating up 13 bad guys, some bar and line graphs, and some Venn diagrams. They were pretty good.

I created most of the graphics myself, except for the Einstein and Pink Panther figures. Well, these two are old and nice, so let's hope they don't mind my using their lovely images.

In addition to MS Word and Paint, I have also used Desmos to create some nice graphs. Thank you Desmos. I am sure you are as cool as I am. I will buy you a drink if you come to town. Look me up on the Internet, I am easy to find, there is only one Tityik Wong in the world.

Index

FINITE MATHEMATICS
Concepts & Applications

From the window of my high-rise condo I can watch the traffic on the I-15 freeway below. I see many cars moving but there is no traffic jam. The Ma-Phone I bought from Mars last summer can detect and record the speed of every single car that passes a fixed line across the freeway. The flow of traffic is smooth, which means most cars are moving at about the same speed. Once in a while I see a bald old man in a Venom Model 8 zooming by everybody, or a long-haired young hippie

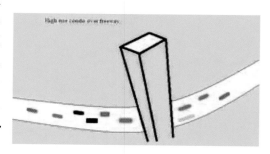

on a square-wheeled homemade motor cycle driving very slowly in the emergency lane while juggling. But overall, most people drive no*rma*lly. After one hour of watching my Ma-Phone shows the following speed distribution:

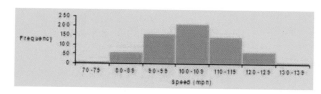

The day is still young and I have nothing to do so I decide to watch for another hour. After that the following speed distribution emerges:

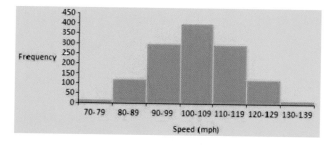

I see a bell-shaped curve forming and ask my Me-Professor from Mercury what is happening. My Me-Professor replies: "It seems like the shape of the distribution is approaching a normal curve, dude! I bet you don't know what a normal curve is, do you?"

  I throw the professor into the bathtub. After that I use the Ve-Device I imported from Venus to look up the term "normal distribution".

Made in the USA
Las Vegas, NV
18 January 2022